S0-DTP-735

# Electrical Engineering Materials

PRENTICE-HALL ELECTRICAL ENGINEERING SERIES

W. L. EVERITT, Ph.D., *Editor*

ANNER *Elements of Television Systems*
BALABANIAN *Network Synthesis*
BENEDICT *Introduction to Industrial Electronics*
DAVIS AND WEED *Industrial Electronic Engineering*
DEKKER *Electrical Engineering Materials*
FETT *Feedback Control Systems*
FICH *Transient Analysis in Electrical Engineering*
FICH AND POTTER *Theory of A-C Circuits*
GOLDMAN *Information Theory*
GOLDMAN *Transformation Calculus and Electrical Transients*
HERSHBERGER *Principles of Communication Systems*
JORDAN *Electromagnetic Waves and Radiating Systems*
LO, ENDRES, ZAWELS, WALDHAUER AND CHENG *Transistor Electronics*
MARTIN *Electronic Circuits*
MARTIN *Physical Basis for Electrical Engineering*
MARTIN *Ultrahigh Frequency Engineering*
MOSKOWITZ AND RACKER *Pulse Techniques*
NIXON *Principles of Automatic Controls*
PUMPHREY *Electrical Engineering*, 2d ed.
PUMPHREY *Fundamentals of Electrical Engineering*
REED *Electric Network Synthesis*
RIDEOUT *Active Networks*
RYDER *Electronic Engineering Principles*, 2d ed.
RYDER *Electronic Fundamentals and Applications*
RYDER *Networks, Lines, and Fields*, 2d ed.
SHEDD *Fundamentals of Electromagnetic Waves*
SKRODER AND HELM *Circuit Analysis by Laboratory Methods*, 2d ed.
STOUT *Basic Electrical Measurements*
VAIL *Circuits in Electrical Engineering*
VAN DER ZIEL *Noise*
VAN DER ZIEL *Solid State Physical Electronics*
VAN VALKENBURG *Network Analysis*
VON TERSCH AND SWAGO *Recurrent Electrical Transients*
WARD *Introduction to Electrical Engineering*, 2d ed.

# ELECTRICAL ENGINEERING MATERIALS

ADRIANUS J. DEKKER

PROFESSOR, DEPARTMENT OF ELECTRICAL ENGINEERING
INSTITUTE OF TECHNOLOGY
UNIVERSITY OF MINNESOTA

*Englewood Cliffs, N. J.*
PRENTICE-HALL, INC.
1959

*Library of Congress Catalog Card Number: 58-8168*

PRINTED IN THE UNITED STATES OF AMERICA
24704

# Preface

It might be said justifiably that the curriculum offered in most electrical engineering departments has changed a good deal during the last five or ten years. In our own department, for example, there has been an increasing emphasis on the teaching of fundamental concepts rather than more specialized subjects. This is true particularly of the first four years of our five-year undergraduate curriculum, the fifth year being reserved for the discussion of the specialized areas. Running parallel with this tendency there has been an increasing emphasis on a sound background in physics and mathematics.

Another aspect of the changes in the curriculum involves the increasing importance of the science of materials, which has led to a number of new devices used in present-day electrical engineering, and which will probably become even more important in the future. In our own department this has resulted in a gradual filtering-down of subjects such as electron physics and "molecular engineering" from the graduate level to the undergraduate level. Thus, a few years ago an elective course in the fifth year of the undergraduate program was introduced in which the operation of solid state devices is discussed along the lines of A. van der Ziel's book *Solid State Physical Electronics* (Prentice-Hall, Inc., 1957).

More recently, the faculty of our department decided to go one step further in this direction by offering a new course to be taken by all electrical engineering students in their fourth year. The course was to run for only one quarter, at least to begin with, and its purpose was to introduce the students to the physical interpretation of the dielectric, magnetic, and conductive properties of materials without entering into the actual applications. This book is a modest attempt to provide material for a course of this kind. It is limited in scope, but the subjects discussed would seem particularly suitable for a course aimed at giving the students some idea of the methods and models employed in the study of materials of interest to the electrical engineer. Also, I believe that it is more useful for a student to have absorbed a certain amount of knowledge about a limited number of subjects than to have been exposed to a great variety in a hurry.

Since most undergraduate electrical engineers have no working knowledge of wave mechanics, no attempt has been made to introduce quantum mechanical concepts, except in a passing manner. This may disappoint some of my colleagues, who will point out that one can introduce wave mechanics by qualitative arguments. However, I am not convinced that the students will benefit greatly from such arguments at this level unless they have acquired a certain degree of maturity in handling classical problems. Thus, the models used in this book are essentially classical or semiclassical. I feel that the lack of rigor implied by these models is outweighed by their usefulness in providing the student with a reasonable amount of insight into the physical mechanisms which underlie the properties of materials. I have also found that these models provide good exercise for the student, to which he can apply his knowledge of elementary field theory.

Although it is not necessary to adhere to the order in which the subjects are discussed in this book, it would seem desirable to have dielectrics precede magnetics; on the other hand, one may well discuss all or part of the last three chapters on conduction in metals and semiconductors before one deals with dielectric and magnetic properties.

A list of general references is given at the beginning of this book, whereas references to specialized topics can be found at the end of each chapter. In general, I have limited the references to representative books or review articles. A set of problems has been given at the end of each chapter. In a number of cases these problems are intended to supplement the text. I have refrained from giving many problems which merely require application of the slide rule. However, a number of numerical problems have been included to give the student a feeling for the order of magnitude of the quantities which enter into the discussion. Answers to the problems are provided at the end of the book; a table of frequently occurring physical constants may be found at the beginning. Throughout this book, the mks system of units has been used.

I wish to express my appreciation to Dr. W. G. Shepherd for his encouragement before and during the preparation of the manuscript, and to Dr. K. M. van Vliet for valuable discussions and for reading a large part of the manuscript.

A. J. Dekker

# Contents

# General References

A. J. Dekker, *Solid State Physics*, Prentice-Hall, Englewood Cliffs, N. J., 1957.

C. Kittel, *Introduction to Solid State Physics*, Wiley, New York, 2nd ed., 1956.

T. L. Martin, Jr., *Physical Basis for Electrical Engineering*, Prentice-Hall, Englewood Cliffs, N. J., 1957.

D. C. Peaslee and H. Mueller, *Elements of Atomic Physics*, Prentice-Hall, Englewood Cliffs, N. J., 1955.

M. J. Sinnott, *The Solid State for Engineers*, Wiley, New York, 1958.

A. van der Ziel, *Solid State Physical Electronics*, Prentice-Hall, Englewood Cliffs, New Jersey, 1957.

C. Zwikker, *Physical Properties of Solid Materials*, Interscience, New York, 1954.

J. E. Goldman, ed., *The Science of Engineering Materials*, Wiley, New York, 1957.

# Approximate Values of Physical Constants

| | |
|---|---|
| Avogadro's number | $6.0254 \times 10^{23}$ per gram molecule |
| Boltzmann's constant, $k$ | $1.380 \times 10^{-23}$ joule degree$^{-1}$ |
| Electric conversion factor, $\epsilon_0$ | $8.854 \times 10^{-12}$ farad meter$^{-1}$ |
| Electronic charge, $e$ | $-1.601 \times 10^{-19}$ coulomb |
| Electron rest mass, $m$ | $9.107 \times 10^{-31}$ kilogram |
| Loschmidt's number | $2.687 \times 10^{25}$ meter$^{-3}$ |
| Magnetic conversion factor, $\mu_0$ | $4\pi \times 10^{-7} = 1.257 \times 10^{-6}$ henry meter$^{-1}$ |
| Planck's constant, $h$ | $6.624 \times 10^{-34}$ joule second |
| Proton rest mass | $1.672 \times 10^{-27}$ kilogram |
| Velocity of light, $c$ | $2.998 \times 10^{8}$ meter second$^{-1}$ |
| 1 Bohr Magneton, $\beta = eh/4\pi m$ | $9.27 \times 10^{-24}$ ampere meter$^{2}$ |
| 1 Debye unit | $3.33 \times 10^{-30}$ coulomb meter |

# I

# Atoms and Aggregates of Atoms

## 1.1 Introduction

The physical behavior of a given material may be characterized by a set of *macroscopic*, measurable quantities, such as the electrical conductivity of the material, its coefficient of expansion, its magnetic permeability, its dielectric constant, etc. In general, these quantities are functions of externally variable parameters, such as temperature, pressure, frequency of the applied field, etc. The functional relationships between the characteristic quantities and the variable parameters can be established from experimental results, and constitute an important part of our technical and scientific knowledge. In this book we shall be concerned with problems which arise when one asks why a certain functional relationship between a characteristic quantity and a parameter exists. We shall try to answer such a question in terms of the properties of the atoms which constitute the material; i.e. we shall accept the idea that materials consist of atoms, and that atoms consist of nuclei and electrons, and shall attempt to derive the observed relationships in terms of our knowledge of atoms. Actually we shall, in most cases, assume a much simpler "*model*" for the atomic constituents than is justified in terms of our present-day knowledge of atoms. The reason for this is that in most cases the calculations involved in arriving at a certain relationship would be too complicated if they were attempted on the basis of first principles. For example, it is impossible to calculate exactly the dielectric constant of a material. On the other hand, we can learn a great deal about the behavior of dielectrics if we are willing to accept certain simplifications concerning the structure and properties of the atoms. Simplified atomic models will thus play an important role in the discussions in this book. The reader should always realize that the results of calculations pertaining to such models cannot be expected to provide exact numbers. The main purpose of the model is that it can provide the correct functional relationship between certain quantities, and thereby provide insight into the essential mechanism which determines such a relationship. On the basis of this understanding predictions can frequently be made concerning the properties of a large group of materials. In other words, the atomic models unify our understanding of the properties of materials and as such prove their scientific usefulness.

In this book we shall discuss only a limited number of properties. In particular we shall be concerned with the dielectric and magnetic properties of materials used in electrical engineering, and with the mechanism of electrical conductivity. The materials under discussion are in most cases crystalline solid materials. It is desirable to realize from the beginning that the properties of solids are not simply given by the sum total of the properties of the atoms which make the solid. In fact, a property such as the electrical conductivity of a solid can be understood only as a consequence of the strong interaction between the atoms. This interaction may change the properties of the individual atoms to such an extent that the properties of the solid may be completely different from those of the separate atoms.

In this first chapter we shall present some material which will be used in later chapters. It seems in order, for example, to make some remarks about atoms and their structure, about the arrangement of atoms in solids, and about the interaction between atoms in solids. Since this is not a book on the physics of atoms or a general book on the physics of solids, it may suffice to give here only the most essential elements of these subjects. In the remaining chapters, these subjects will be extended as the need arises.

## 1.2 The hydrogen atom according to the old and new quantum mechanics

There is a vast amount of experimental evidence which shows that an atom consists of a positively charged nucleus and a number of negatively charged electrons which revolve about the nucleus. The nucleus may be considered to be built up of a number of neutral particles (neutrons) and a number of positively charged protons. The charge of the nucleus is thus determined by the number of protons it contains, $Z$; the charge per proton is $e = 1.60 \times 10^{-19}$ coulomb. In a neutral atom, the number of electrons is equal to $Z$, the charge carried by each electron being equal to $-e$. The mass of an electron is $m = 9.107 \times 10^{-31}$ kg, and is approximately 1836 times smaller than that of a proton or a neutron. Hence, practically the whole mass of an atom is concentrated in the nucleus.

The size of an atom is not a well-defined quantity, but may be said to be of the order of 1 angstrom $= 10^{-10}$ m. The classical radius of an electron and of a nucleus, on the other hand, is only of the order of $10^{-15}$ m; consequently, in terms of a classical representation of electrons, an atom is essentially "empty." In terms of the wave mechanical interpretation, however, it is better to think of the electrons revolving about the nucleus as a *continuous charge distribution.* The shape of the charge distribution is determined by the state of motion of the electron. The difference between a semi-classical representation of an electron and the wave mechanical representation may be illustrated by discussing briefly the structure of the hydrogen atom according to the old quantum theory of Bohr (1913), and by comparing this result with that obtained on the basis of wave mechanics (Schrödinger, Heisenberg, 1924).

A hydrogen atom consists of an electron moving in the field of a proton. Assuming the electron revolves as a point-like particle in a circular orbit of radius $r$ around the proton, the stability of the orbit requires equilibrium

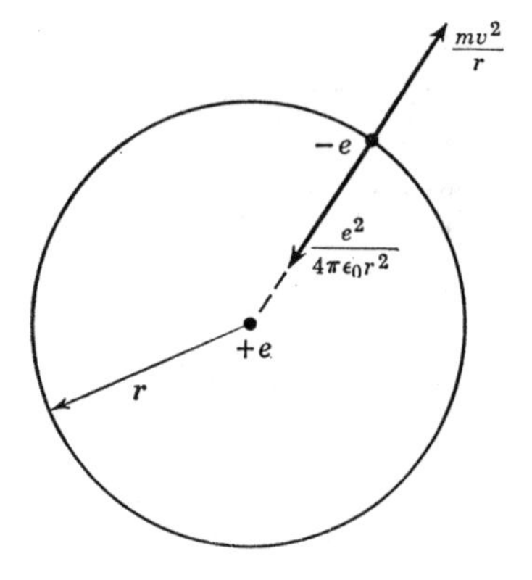


**Fig. 1.1.** Illustrating the forces corresponding to a circular orbit of an electron in a hydrogen atom.

between the attractive Coulomb force on the electron and the centrifugal force. Hence, we must require with reference to Fig. 1.1

$$\frac{mv^2}{r} = \frac{e^2}{4\pi\epsilon_0 r^2} \tag{1.1}$$

Here, $v$ is the velocity of the electron in the orbit and $\epsilon_0 = 8.854 \times 10^{-12}$ farad m$^{-1}$. The total energy of the electron, $W$, in this state of motion is equal to the kinetic energy, $(1/2)mv^2$, plus the potential energy due to the Coulomb field of the proton. Defining the potential energy of the electron for $r \to \infty$ as zero, we thus may write

$$W = \frac{1}{2}mv^2 - \frac{e^2}{4\pi\epsilon_0 r} \tag{1.2}$$

Substituting for $(1/2)mv^2$ from (1.1) into (1.2) we obtain

$$W = -\frac{e^2}{8\pi\epsilon_0 r} \tag{1.3}$$

The minus sign indicates that the electron has less energy in the orbit $r$ than it would have if $r$ were infinite. In other words, the electron is bound in the field of the nucleus, the energy required to take it away from the nucleus being equal to the positive quantity $-W$. Up to this point, the treatment falls completely within the realm of classical physics. However, it is well known that this classical atomic model is unstable because the energy would decrease continually as a result of emission of electromagnetic radiation; ultimately, the electron would spiral into the nucleus. In order to retain the stability of the orbit, Bohr postulated a *quantum condition* on the motion of the electron by assuming that only those circular orbits are stable for which the angular momentum is equal to an integer times $h/2\pi$, where $h = 6.62 \times 10^{-34}$ joule sec represents Planck's constant. Mathematically this quantum condition for a circular orbit reads

$$mvr = nh/2\pi \quad \text{where} \quad n = 1, 2, 3, \ldots \tag{1.4}$$

On the basis of this postulate one finds a set of energy levels $W_n$ which the orbiting electron may assume. Thus, by substituting for $v^2$ from (1.4) into (1.1) one obtains for the possible radii of the circular orbits

$$r_n = \frac{\epsilon_0 h^2}{\pi m e^2} n^2 = 5.29 \times 10^{-10}\, n^2 \text{ meter} \tag{1.5}$$

Thus, the smallest radius the electron orbit can assume is 5.29 angstroms; the next possible radii are 4, 9, 16, etc. times as large. In accordance with (1.3) this means that the energy of the electron can accept only a *series of discrete values*, $W_n$, given by

$$W_n = -\frac{me^4}{8\epsilon_0^2 h^2}\frac{1}{n^2} = -\frac{13.6}{n^2} \text{ electron volts} \tag{1.6}$$

(One electron volt $= 1$ ev $= 1.6 \times 10^{-19}$ joule.) Thus, in its lowest state (the *ground state*) the electron is bound to the field of the proton to the extent of 13.6 ev, i.e. it takes an energy of 13.6 ev to ionize the hydrogen atom. The energy levels are represented schematically in Fig. 1.2. As explained in some detail in courses on atomic physics, this energy level diagram agrees satisfactorily with certain parts of the emission and absorption spectra of

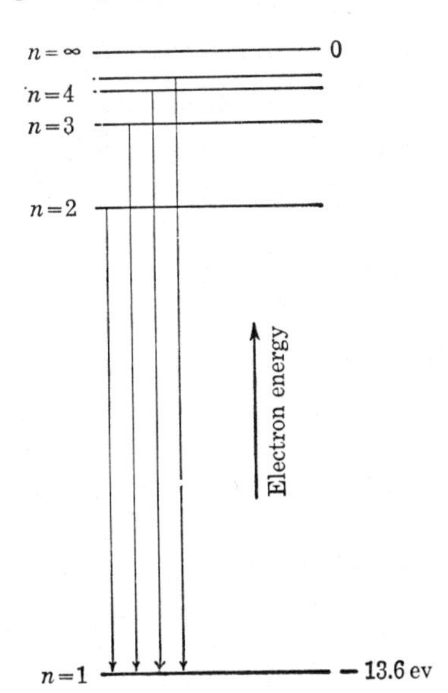


**Fig. 1.2.** Energy levels of an electron in a hydrogen atom; the levels for $n > 5$ have not been indicated. The vertical arrows indicate transitions from higher levels to the ground state; these transitions correspond to emission of electromagnetic radiation.

hydrogen, if one makes the further postulate that a transition of the electron from an energy level $W_{n_1}$ to another level $W_{n_2}$ is associated with the emission or absorption of electromagnetic radiation of a frequency $\nu$ such that

$$h\nu = |W_{n_1} - W_{n_2}| \tag{1.7}$$

The *ad hoc* postulate of Bohr expressed by equation (1.4) obtains a definite physical meaning when considered in the light of *wave mechanics.* We shall not enter into this subject here, and it may suffice to make some general remarks. In wave mechanics, particles are described by waves; the intensity of the waves in a certain volume element in space is interpreted as representing the probability of finding the particle in this volume element. Thus, wave mechanics is *statistical* in the sense that it does not give a definite answer as to where the particle "is" at a certain instant; it tells us only what the *probability* is of finding it in a certain small volume element in space. Consequently, in describing an electron in an atom in terms of wave mechanics one ends up with a certain *charge distribution* of a "smeared-out" electron. The *wave function,* which represents these waves and hence determines the charge distribution associated with the electron, satisfies the so-called *Schrödinger wave equation.* This equation is a partial differential equation which replaces the classical Newtonian equations of motion. When applied to the problem of an electron moving in the field of a proton, it turns out that physically acceptable solutions for the wave function exist only for specific integer values of three *quantum numbers:*

the *principal* quantum number $n = 1, 2, 3, \ldots$ (1.8)

the *angular momentum* quantum number $l = 0, 1, \ldots, (n - 1)$ (1.9)

the *magnetic* quantum number $m_l = l, (l - 1) \ldots, -(l - 1), -l$ (1.10)

In quantum mechanics then, the quantum numbers arise as a natural consequence of the *wave nature of matter;* for a discussion of the experimental evidence which supports this notion, we refer the reader to textbooks on the subject. The principal quantum number fulfills the same role as the quantum number $n$ in the theory of Bohr; i.e., the energy levels obtained from wave mechanics are the same as those given by the Bohr formula (1.6). As a result of the wave nature of the electrons, however, the interpretation of the motion of the electron in the ground state in terms of a circular orbit in the Bohr theory is replaced by an interpretation in terms of a charge distribution in the wave mechanical theory. For example, the charge density associated with an electron moving about the nucleus in the ground state of a hydrogen atom is given by

$$\rho(r) = -(e/\pi r_1^3)e^{-2r/r_1} \tag{1.11}$$

where $r_1$ is equal to 5.29 angstroms, i.e. $r_1$ is equal to the radius of the first Bohr orbit; see Fig. 1.3(a). The total charge corresponding to this charge

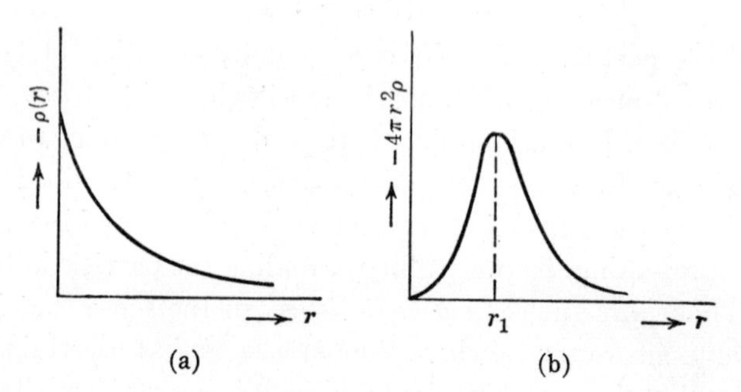


**Fig. 1.3.** A schematic representation of the electronic charge density $\rho(r)$ of a hydrogen atom in the ground state is given in (a). In (b) the integrand of expression (1.12) is represented schematically; $r_1$ is the first Bohr radius of the hydrogen atom.

distribution is, of course, equal to the electronic charge; i.e.

$$\int_{r=0}^{\infty} 4\pi r^2 \rho(r)\, dr = -e \tag{1.12}$$

as may readily be verified by the reader. The amount of charge contained in a shell between two concentric spheres of radii $r$ and $r + dr$ is presumably given by the integrand in expression (1.12). The integrand has its maximum value for $r = r_1$, as indicated in Fig. 1.3(b). Hence, in the wave mechanical theory, the maximum of the charge distribution in the ground state occurs for a distance from the nucleus equal to the first Bohr radius.

The quantum number $l$ determines the angular momentum of the electron, whereas $m_l$ determines the *component* of the angular momentum along a prescribed direction, which may be, for example, the direction of an external magnetic field. The physical meaning of these quantum numbers will be discussed further in the chapter dealing with the magnetic properties of atoms.

## 1.3 Nomenclature pertaining to electronic states

In the preceding section it was noted that the state of motion of the electron in a hydrogen atom can be described by three quantum numbers $n$, $l$ and $m_l$, and a set of these numbers is said to define the *state* of the electron. In the lowest energy level ($n = 1$) the quantum numbers $l$ and $m_l$ must both be zero in accordance with the rules (1.9) and (1.10). Thus, the ground state of the hydrogen atom is defined by $n = 1$, $l = 0$ and $m_l = 0$. If the electron is in a higher energy level, say in the level corresponding to $n = 2$, various states are possible. In fact, by applying the rules (1.9) and (1.10) we find the possible states

$$\begin{array}{lll} n = 2 & l = 0 & m_l = 0 \\ n = 2 & l = 1 & m_l = 1 \\ n = 2 & l = 1 & m_l = 0 \\ n = 2 & l = 1 & m_l = -1 \end{array} \tag{1.13}$$

Each of these states corresponds to a particular charge distribution of the "smeared-out" electron in wave mechanics. In atomic physics, states with a particular $l$-value have a particular name. Thus, a state with $l = 0$ is called an "$s$-state"; a state with $l = 1$ is called a "$p$-state," etc. These names are derived from the nomenclature used in the classification of spectral lines of atoms. We give here the names for the states corresponding to various values of $l$:

$$\begin{array}{lcccccc} & l = 0 & 1 & 2 & 3 & 4 & \ldots \\ \text{name the} & & & & & & \\ & s & p & d & f & g & \ldots \text{-states} \end{array} \tag{1.14}$$

We may now raise the question: for a given value of the principal quantum number $n$, how many electronic states are possible? Presumably, what we are asking for is the number of different sets of $n$, $l$, $m_l$ values which exist for a given value of $n$, assuming that the quantum numbers satisfy the rules (1.9) and (1.10). To answer this question, we first note that according to (1.10) there are $(2l + 1)$ possible values of $m_l$ for a given value of $l$. Furthermore, according to (1.9) $l$ can accept a total of $n$ different values for a given value of $n$. Hence, the total number of states corresponding to a given value of $n$ is equal to

$$\sum_{l=0}^{l=n-1} (2l + 1) = 1 + 3 + \ldots [2(n - 1) + 1] = n^2 \qquad (1.15)$$

Thus, for $n = 2$, there are $2^2 = 4$ different states, which is confirmed by the result in (1.13); for $n = 3$, there are 9 different states, etc. It is emphasized that an *energy level is not equivalent with an electronic state;* an energy level is determined by the value of $n$, and such a level thus corresponds to $n^2$ states. It should be mentioned here that actually the energy of an electron is also determined to some extent by the quantum numbers $l$ and $m_l$; however, the differences in energy between an electron in the state, $n$, $l_1$, $m_{l1}$, and an electron in the state $n$, $l_2$, $m_{l2}$ is very small compared to the energy difference between two states of different $n$-values.

The group of states corresponding to a given value of the principal quantum number $n$ is referred to as a *shell* of electrons. Thus, the states corresponding to $n = 1$ form the $K$-shell; those corresponding to $n = 2$ form the $L$-shell, etc. Hence,

$$n = 1 \quad 2 \quad 3 \quad 4 \quad 5 \quad \ldots$$

correspond to the

$$K \quad L \quad M \quad N \quad O \quad \ldots \quad \text{-shells}$$

## 1.4 The electron configuration of atoms

In an atom containing more than one electron, the nomenclature given for the electronic states is retained. In determining the states of the electrons in a many-electron atom, the *Pauli exclusion principle* must be introduced. This principle says that *a given quantum state determined by three quantum numbers $n$, $l$, $m_l$ can be occupied by not more than two electrons.** For example, the $K$-shell of an atom corresponds to $n = 1$ and thus contains only 1 state, viz. $n = 1$, $l = 0$ and $m = 0$. According to the Pauli exclusion principle, there can be no more than 2 electrons in the $K$-shell. Similarly, the $L$-shell, corresponding to $n = 2$ has 4 different sets of values $n$, $l$, $m_l$ and hence can contain no more than $2 \times 4 = 8$ electrons. In gen-

* The factor 2 arises from the *spin quantum number*, $s$, which can accept two possible values; for further details, see section 4.8.

eral, the electron shell corresponding to the principal quantum number $n$ can contain no more than $2n^2$ electrons.

These rules are important for the interpretation of the periodic system of the elements in terms of the electron configuration of the atoms. For

**Table 1.1.** THE ELECTRON CONFIGURATION OF THE FIRST 36 ELEMENTS

| | | $K$ $n = 1$ | $L$ $n = 2$ | | $M$ $n = 3$ | | | $N$ $n = 4$ | | | |
|---|---|---|---|---|---|---|---|---|---|---|---|
| Atomic Number $Z$ | Element | $l = 0$ $s$ | $l = 0$ $s$ | $l = 1$ $p$ | $l = 0$ $s$ | $l = 1$ $p$ | $l = 2$ $d$ | $l = 0$ $s$ | $l = 1$ $p$ | $l = 2$ $d$ | $l = 3$ $f$ |
| 1 | H | 1 | | | | | | | | | |
| 2 | He | 2 | | | | | | | | | |
| 3 | Li | 2 | 1 | | | | | | | | |
| 4 | Be | 2 | 2 | | | | | | | | |
| 5 | B | 2 | 2 | 1 | | | | | | | |
| 6 | C | 2 | 2 | 2 | | | | | | | |
| 7 | N | 2 | 2 | 3 | | | | | | | |
| 8 | O | 2 | 2 | 4 | | | | | | | |
| 9 | F | 2 | 2 | 5 | | | | | | | |
| 10 | Ne | 2 | 2 | 6 | | | | | | | |
| 11 | Na | 2 | 2 | 6 | 1 | | | | | | |
| 12 | Mg | 2 | 2 | 6 | 2 | | | | | | |
| 13 | Al | 2 | 2 | 6 | 2 | 1 | | | | | |
| 14 | Si | 2 | 2 | 6 | 2 | 2 | | | | | |
| 15 | P | 2 | 2 | 6 | 2 | 3 | | | | | |
| 16 | S | 2 | 2 | 6 | 2 | 4 | | | | | |
| 17 | Cl | 2 | 2 | 6 | 2 | 5 | | | | | |
| 18 | A | 2 | 2 | 6 | 2 | 6 | | | | | |
| 19 | K | 2 | 2 | 6 | 2 | 6 | | 1 | | | |
| 20 | Ca | 2 | 2 | 6 | 2 | 6 | | 2 | | | |
| 21 | Sc | 2 | 2 | 6 | 2 | 6 | 1 | 2 | | | |
| 22 | Ti | 2 | 2 | 6 | 2 | 6 | 2 | 2 | | | |
| 23 | V | 2 | 2 | 6 | 2 | 6 | 3 | 2 | | | |
| 24 | Cr | 2 | 2 | 6 | 2 | 6 | 5 | 1 | | | |
| 25 | Mn | 2 | 2 | 6 | 2 | 6 | 5 | 2 | | | |
| 26 | Fe | 2 | 2 | 6 | 2 | 6 | 6 | 2 | | | |
| 27 | Co | 2 | 2 | 6 | 2 | 6 | 7 | 2 | | | |
| 28 | Ni | 2 | 2 | 6 | 2 | 6 | 8 | 2 | | | |
| 29 | Cu | 2 | 2 | 6 | 2 | 6 | 10 | 1 | | | |
| 30 | Zn | 2 | 2 | 6 | 2 | 6 | 10 | 2 | | | |
| 31 | Ga | 2 | 2 | 6 | 2 | 6 | 10 | 2 | 1 | | |
| 32 | Ge | 2 | 2 | 6 | 2 | 6 | 10 | 2 | 2 | | |
| 33 | As | 2 | 2 | 6 | 2 | 6 | 10 | 2 | 3 | | |
| 34 | Se | 2 | 2 | 6 | 2 | 6 | 10 | 2 | 4 | | |
| 35 | Br | 2 | 2 | 6 | 2 | 6 | 10 | 2 | 5 | | |
| 36 | Kr | 2 | 2 | 6 | 2 | 6 | 10 | 2 | 6 | | |

further reference we give in Table 1.1 the electron configuration of a number of atoms. This table shows that up to element 19 (potassium), the filling of the electronic states is completely regular in the sense that higher levels are not filled until lower levels are occupied by the maximum allowable number of electrons. In the element potassium, however, we note that the 4*s*-level contains one electron while the 3*d*-states are still unoccupied (4*s* means $n = 4$, $l = 0$; 3*d* means $n = 3$, $l = 2$). This situation of incompletely filled 3*d*-states persists until in element 29 (copper) the 3*d*-states are occupied by the maximum number of electrons, viz. 10. A group of elements for which parts of an inner shell are not occupied by electrons is called a group of *transition elements*. The particular group for which the 3*d*-states are partly empty is called the iron group. In the chapter dealing with the magnetic properties of materials we shall see that these properties are determined to a large extent by the incompletely filled inner states.

Our knowledge concerning the electron configuration of atoms has contributed a great deal to the understanding of the periodicity of the chemical properties as expressed by the arrangement of the elements in the *periodic table*. The reason is that the *chemical properties of atoms are determined mainly by the outer electron configuration*. Thus, elements such as the alkali metals (Li, Na, K, Rb and Cs) all have one outer electron, and all behave chemically in a similar fashion. The reason for the important role played by the outer electrons in determining the chemical properties of atoms is readily understood. When an atom A is brought close to another atom B, the electrons in the A atom will be subjected to forces which were not present in the absence of B. However, the inner electrons in the A atom are under influence of the strong Coulomb field produced by the nucleus of atom A and hence the perturbing fields produced by atom B are of much less consequence than they are for the more weakly bound outer electrons. The perturbing fields may thus distort the charge distribution of the outer electrons of A and B atoms to such an extent that a *chemical bond* may result. The nature of the chemical bond will be discussed briefly in the next section.

## 1.5 The nature of the chemical bond and the classification of solids

Although we do not intend to discuss the nature of the various kinds of chemical bonds in any detail here, a few general remarks may be in order because the type of bonding between atoms determines to a large extent the electrical and other physical properties of solid materials. From a purely phenomenological point of view, we may argue that in a solid material there are two types of forces acting between the atoms: (a) *attractive forces*, which keep the atoms together so as to form a solid, (b) *repulsive forces* which become noticeable when one attempts to compress a solid. These arguments apply as well to liquids and, in fact, to single molecules. It is important to realize, however, that the mere existence of attractive and repulsive forces between atoms is not sufficient to guarantee the formation of a stable chemical bond. This may be illustrated by considering the following model: Suppose two atoms A and B exert attractive and repulsive forces on each other such that the potential energy of B in the field of A is given by

$$W(r) = -\frac{\alpha}{r^n} + \frac{\beta}{r^m} \tag{1.16}$$

where $r$ is the distance between the centers of the atoms; $n$ and $m$ are arbitrary positive powers, and $\alpha$ and $\beta$ are positive constants which determine the strength of the attractive and repulsive forces, respectively. The zero of energy is chosen such that for $r \to \infty$, the potential energy of the particles in each other's field vanishes. For the moment we are not concerned about the physical origin of these forces. The question is, will these particles form a stable chemical compound or not? The answer is, that this will only be the case if the function $W(r)$ exhibits a minimum for a finite value of $r$, as illustrated in Fig. 1.4. If such a minimum exists, the two

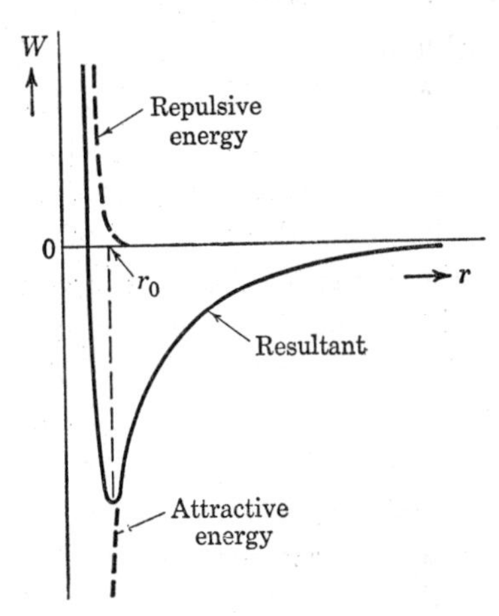


**Fig. 1.4.** Representation of the potential energy between two atoms as a function of their distance from one another. The stable molecule corresponds to the separation $r_0$, the energy then being a minimum.

atoms will form a stable compound with a distance between them equal to the value $r_0$ for which the minimum in $W(r)$ occurs; the energy required to dissociate the molecule is then equal to the positive quantity $-W(r_0)$. In order for $W(r)$ to exhibit a minimum, the powers $n$ and $m$ in (1.16) must satisfy a certain condition, viz.

$$m > n \tag{1.17}$$

In other words, the attractive forces must vary more slowly with $r$ than the repulsive forces. Qualitatively, this is evident from the shape of the attractive and repulsive energy in Fig. 1.4. Mathematically, this can be shown as follows: If $W(r)$ exhibits a minimum for $r = r_0$, then we must require

$$(dW/dr)_{r=r_0} = 0 \quad \text{or} \quad r_0^{m-n} = (m/n)(\beta/\alpha) \tag{1.18}$$

and at the same time

$$\left(\frac{d^2W}{dr^2}\right)_{r=r_0} = -\frac{n(n+1)\alpha}{r_0^{n+2}} + \frac{m(m+1)\beta}{r_0^{m+2}} > 0 \tag{1.19}$$

The condition (1.17) then follows immediately by substituting $r_0$ from (1.18) into (1.19).

The forces acting between atoms are of an electrostatic nature and, as mentioned in the preceding section, are determined essentially by the extent to which the wave functions of the outer electrons are perturbed by the presence of other atoms at close proximity. On the basis of the type of chemical bond, solids may be classified as follows:

(i) Ionic crystals (NaCl, KF)
(ii) Valence crystals (diamond, Si, Ge, SiC)
(iii) Metals (Cu, Ag, Fe)
(iv) van der Waals crystals (solid argon, organic crystals)

With reference to each of these classes we shall now make some general remarks concerning the chemical bonds in these materials.

(i) **Ionic crystals.** Ionic crystals are formed by combining two or more kinds of atoms which differ considerably in their tendencies to give off or to accept electrons. For example, when sodium and chlorine atoms are combined, the energy of the solid in which the sodium atoms become positive ions and the chlorine atoms become negative ions is lower than that in which the atoms remained neutral. The outer $3s$ electrons of the sodium atoms are transferred to the chlorine atoms which thereby obtain a stable configuration of 6 electrons in the $3p$-states. Now, the ionization of a free sodium atom requires an energy of 5.1 ev, whereas the energy gained by putting an electron on a neutral Cl atom is 4 ev. One might get the impression that transfer of an electron from sodium to chlorine is unfavorable in the sense that the net energy expended is positive. This is indeed the case if the two atoms are a great distance apart. However, when the $Na^+$ ion and the $Cl^-$ ion are brought together so that their nuclei are separated by only a few angstroms, *energy is gained as a result of the Coulomb attrac-*

*tion between the ions.* Thus, in the solid state, sodium chloride is built up of $Na^+$ and $Cl^-$ ions, rather than of neutral atoms. The Coulomb forces between these ions are mainly responsible for keeping the ions together. Since the electrons in these ions are all rather tightly bound, ionic crystals exhibit in general, no electrical conductivity which can be associated with the motion of electrons. However, at elevated temperatures they do show some electrical conductivity associated with the motion of ions under influence of an electric field.

Ionic solids are formed particularly between elements on the left- and on the right-hand sides of the periodic table. Thus, the alkali halides formed between the alkali metals Li, Na, K, Rb, Cs and the halogens $F_2$, $Cl_2$, $Br_2$, $I_2$ are strongly ionic. On the other hand, a compound such as BaS is probably somewhat less ionic; i.e., the barium atoms do not part completely from their two outer valence electrons. Taking an element such as In from the third column in the periodic table and an element such as Sb from the fifth column gives rise to a compound (in this case indium antimonide) which has very little ionic character at all. These remarks indicate that the classification given above refers to extreme cases and that many solids must be considered as having a chemical bond which lies somewhere between groups (i) and (ii), for example.

Many truly ionic crystals, such as the alkali halides, are transparent for visible light. This property can also be explained in terms of the electron configuration of the material.

(ii) **Valence crystals.** In valence crystals such as diamond, silicon and germanium, the atoms remain neutral. This is not surprising for elements because all atoms are equivalent and there is no reason to assume that some would be ionized while others were not. In a case such as silicon carbide, the atoms are still predominantly neutral, even though they are of different kinds. However, in the case of silicon carbide there may be a slight ionicity involved in the bond because the atoms are different. In a true valence crystal, the binding between the atoms is accomplished by the *sharing of valence electrons.* This will be discussed further in the chapter on semiconductors. These valence or homopolar bonds can be understood only in terms of a wave mechanical theory. In principle, this type of bond is similar to that which exists between two hydrogen atoms in a hydrogen molecule. Valence bonds can be extremely strong, as witnessed by the hardness of materials such as diamond or carborundum.

(iii) **Metals.** In metals the valence electron wave functions are so strongly perturbed by the presence of neighboring atoms that the sharing of these electrons goes so far as to make them highly mobile. In other words, the *valence electrons in a metal cannot be associated with particular atoms; they belong to all atoms.* There is a resemblance between valence crystals and metals because in both cases valence electrons are shared with other atoms. However, in a valence crystal the valence electrons are shared only between nearest neighbor atoms, whereas in a metal the valence electrons are shared by all atoms. Thus, a metal may be considered to consist of an assembly of positive ions embedded in a sea of negative valence electrons. The attractive forces which keep the atoms together arise mainly as a consequence of the Coulomb attraction between the system of positive ions and the negative charge distribution corresponding to the valence electrons. The similarity between valence bonds and metallic bonds will be discussed further in the chapter dealing with semiconductors. It will be argued there that if one considers the elements in the fourth column of the periodic system in the order: diamond, silicon, germanium, gray tin, and lead, one finds a *gradual transition from an extreme valence bond in diamond to a metallic bond in lead.*

(iv) **van der Waals crystals.** It is well known that the atoms of the rare gases such as helium, argon, and neon are chemically extremely inactive; they form no compounds with other atoms. In other words, the outer electron wave functions are not easily perturbed by the presence of other atoms. This chemical inactivity indicates a high degree of stability of the outer electron shell. This is also the reason why these materials remain in the gaseous state at normal temperatures. At very low temperatures, however, a gas like argon will form a solid and the question arises as to what keeps these chemically inactive atoms together. In terms of a classical picture, these weak attractive forces arise as a consequence of the fact that an electron revolving around a nucleus may be considered to represent a rotating electric dipole. Such a dipole will induce a dipole in a neighboring atom such that a dipole-dipole attraction between the atoms results. The process by which a dipole is induced in an atom by means of an electric field will be discussed in detail in the chapter on dielectrics. From a mathematical analysis of the forces acting between neutral atoms it follows that, besides the dipole-dipole interaction just mentioned, there are higher order interactions of the kind dipole-quadrupole, quadrupole-quadrupole, etc. All these forces together are referred to as *van der Waals forces.*

The same kind of forces act between neutral molecules; i.e., many organic molecules form aggregates in which the molecules are held together by van der Waals forces.

A summary of the classification above is given in Fig. 1.5. Starting in the upper left-hand corner, we find metals in which the valence electrons

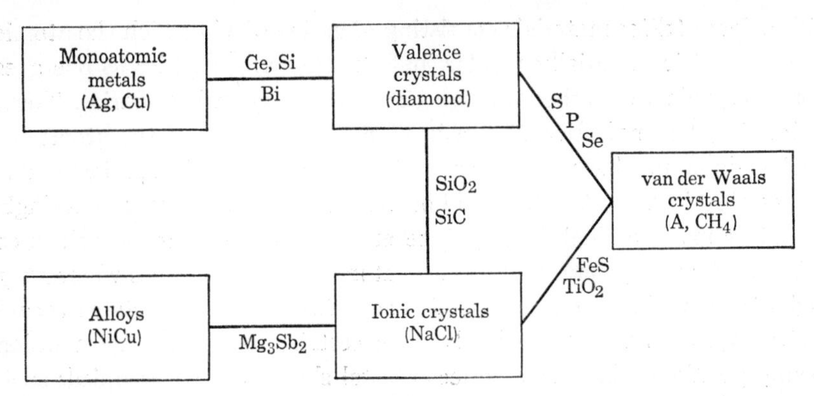


**Fig. 1.5.** Classification of solids; see text.

are shared by all atoms. Related to this group are the valence crystals in which the valence electrons are shared by nearest neighbors. Intermediate, between these two cases are, at least at room temperature, the semiconducting elements such as Ge and Si. At absolute zero, Ge and Si belong in the class of true valence crystals, as we shall see in the chapter on semiconductors. There are also cases which may be considered intermediate between valence and van der Waals solids, such as sulfur, phosphorus, and selenium. Going now to the lower left-hand corner of Fig. 1.5 we see metallic alloys such as nickel-copper. If two metals A and B differ chemically to an appreciable extent, the A atoms may have the tendency to become positive ions whereas the B atoms may have the tendency to become negative ions. Thus, an alloy of magnesium and antimony forms a rather definite compound of chemical composition $Mg_3Sb_2$, corresponding to $Mg^{2+}$ and $Sb^{3-}$ ions, although it is by no means a purely ionic compound. A case like this is intermediate between an alloy and an ionic crystal. Intermediate between ionic crystals and valence crystals are compounds such as SiC and $SiO_2$; here the bonds are partly ionic and arise partly from the sharing of electrons. Materials such as FeS and $TiO_2$ are intermediate between ionic and van der Waals crystals. Thus, we may have ionic bonds in layers of atoms, the layers being held together partly by van der Waals forces.

## 1.6 Atomic arrangements in solids

Most solids to be discussed in this book are *crystalline;* i.e., the atoms or ions are stacked in a regular manner. The fact that a material is crystalline does not necessarily imply that this regular stacking extends throughout the volume of a macroscopic specimen. In fact, one generally deals with *polycrystalline* materials consisting of grains within which the atomic arrangement is essentially regular, but showing irregularities as one goes from one grain to another through the so-called grain boundaries. These grains may be small, say $10^{-6}$ m in diameter, or large. If the atoms are stacked in a regular manner throughout a macroscopic specimen, one speaks of a *single crystal.* It should be mentioned here that even in a single crystal or inside a single grain, there are always certain irregularities or defects. Thus, in general there will be atoms missing at places where they ought to occur in a perfect crystal; one refers to such defects as *vacant lattice sites.* Similarly, there is always a certain number of atoms which occupy positions which in a perfect crystal should not be occupied; such atoms are referred to as *interstitial atoms.* In many cases these defects are very important in explaining the physical properties of materials. For example, the *ionic conductivity* of the alkali halides is due to the presence of vacant lattice sites; if there were no vacant lattice sites, the ions could not move about and the ionic conductivity would be zero at all temperatures. Similarly, in ionic crystals and in metals *diffusion* of atoms takes place by virtue of the presence of vacant lattice sites or interstitial atoms.

Besides the defects just mentioned, there are, of course, always a certain number of foreign atoms present in a material. These impurities may determine to a large extent certain physical properties of the material. This is most dramatically illustrated in the case of *semiconductors,* where the electrical conductivity may be changed by several orders of magnitude by the addition of a fraction of a percent of certain impurities. This subject will be discussed in some detail in Chapter 6. Then there are defects known as *dislocations;* these defects are responsible for the *plastic deformation* of materials. We shall not discuss this subject here, and the reader is referred to the literature on the subject. From these remarks it is evident that an important part of the study of materials is concerned with the study of lattice imperfections. In this section, however, we are concerned not so much with the imperfections as with the regularity of the atomic arrangements in "perfect" crystals.

The most characteristic property of a crystal is its *periodicity of structure.* By this we mean that a crystal may be considered as a repetition in three dimensions of a certain unit pattern, much as certain types of wallpaper have this property in two dimensions. It is not our intention to go into any details concerning crystal structures because it is not necessary for the kind of discussion given in subsequent chapters. It may suffice to give a few examples. In Fig. 1.6 we have presented the crystal structure of NaCl; this structure is typical of the alkali halides, except for the rubidium and

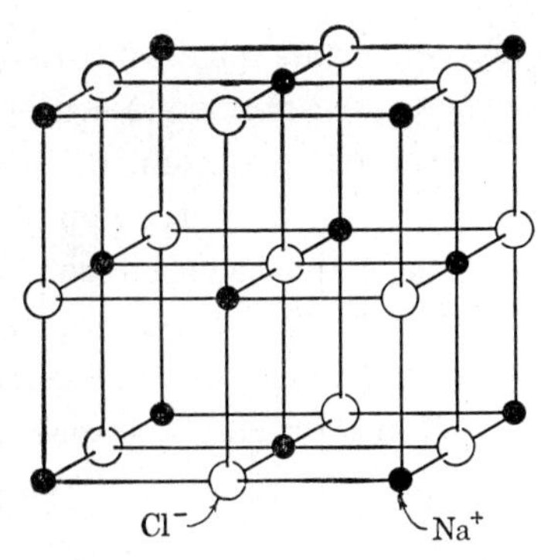


**Fig. 1.6.** The structure of NaCl.

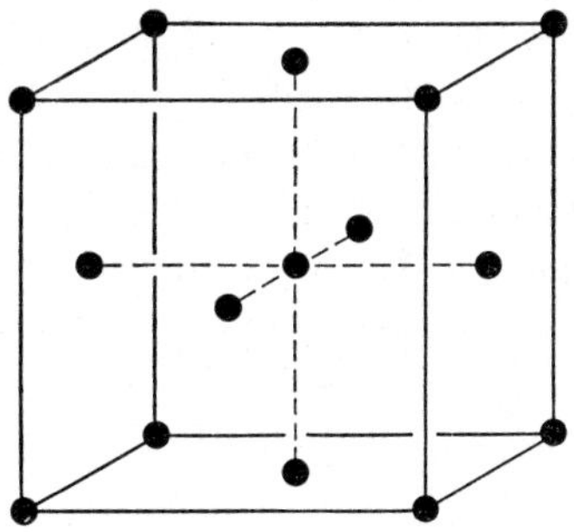

**Fig. 1.7.** The face-centered cubic lattice.

cesium salts. The dots and circles represent the positions of the $Na^+$ and $Cl^-$ nuclei. In an actual crystal neighboring ions "touch" each other and are thus much larger than indicated by the circles and dots (apart from the scaling factor!). It is observed that the $Na^+$ ions occupy the corners of a cube as well as the centers of the cube faces. Such an arrangement is called *face-centered cubic.* Interwoven with this arrangement is a completely equivalent lattice of $Cl^-$ ions. By stacking cubes of this kind in three dimensions; i.e., by repeating the pattern of Fig. 1.6 periodically, a perfect crystal of NaCl would be obtained.

Another example of a face-centered cubic lattice is given in Fig. 1.7, in which all atoms are identical. This is an arrangement found in many metals, such as Cu, Ag, Au, Al, Ni, and a number of others. Several metals crystallize in what is known as a *body-centered cubic structure,* represented in Fig. 1.8; in this case the corners of a cube and the center of the cube are occupied by identical atoms. This structure is found, for example, in Li, Na, K, and Fe.

An example of a somewhat more complicated structure is given in Fig.

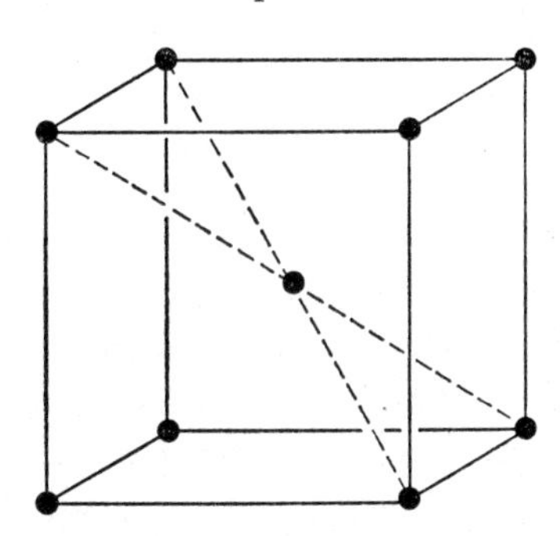

**Fig. 1.8.** The body-centered cubic lattice.

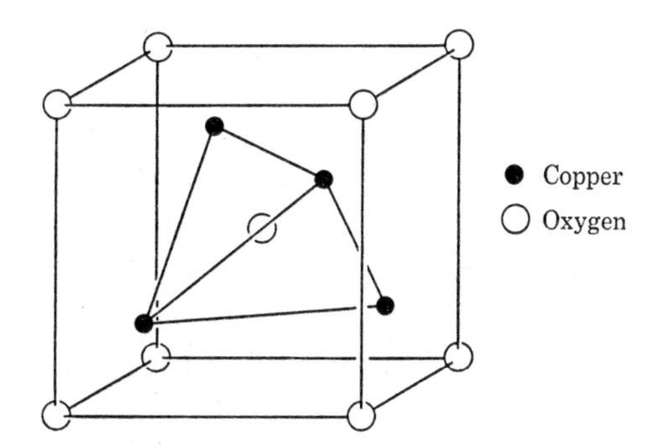


**Fig. 1.9.** The structure of $Cu_2O$.

1.9, for cuprous oxide ($Cu_2O$). Here, the oxygen atoms (or ions) satisfy a body-centered cubic arrangement, whereas the copper atoms are arranged at the corners of a tetrahedron around the central oxygen atom.

One may ask: How does one know that the atoms in a given material are arranged in a particular fashion? The answer is that such information can be obtained from *X-ray* or *electron diffraction patterns*. The principle of X-ray diffraction may be illustrated briefly with reference to Fig. 1.10. The horizontal lines represent planes of atoms in a crystal, the distance

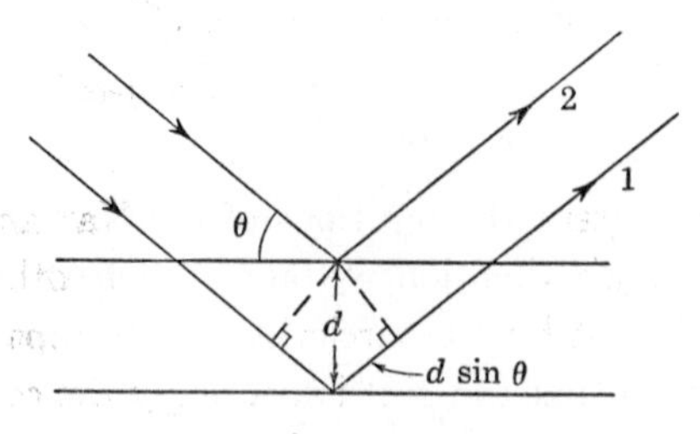


**Fig. 1.10.** Illustrating Bragg reflection of X-rays by a set of equidistant atomic planes.

between these planes being of the order of a few angstroms. Suppose a monochromatic beam of X-rays of a certain wavelength $\lambda$ is incident on the set of planes, the angle between the incident beam and the planes being $\theta$. If there exists a certain relationship between $\lambda$, $\theta$, and the distance $d$ between successive planes, reflection of the X-ray beam may be observed. In general then, for an arbitrary value of $\theta$, assuming $\lambda$ and $d$ to be fixed, no reflection will be observed. The condition which $\theta$ must satisfy in order to observe reflection is evidently that the rays reflected by successive planes are in phase. Thus, if the phase difference between 1 and 2 in Fig. 1.10 is zero, these rays will reinforce each other, and at the same time they will reinforce other rays reflected against deeper lying planes. The condition for reinforcement is clearly that the path difference between 1 and 2 must be equal to an integer times the wavelength of the X-ray beam. Thus,

$$2d \sin \theta = n\lambda \quad \text{where} \quad n = 1, 2, 3, \ldots \tag{1.20}$$

By measuring the angles $\theta$ for which reflection occurs one can thus find the distance between subsequent planes of the set under consideration. From a more detailed analysis one can, in fact, find the arrangement of atoms in the basic unit from which the crystal can be built up. Condition (1.20) is known as the *Bragg condition* for X-ray reflection from a certain set of planes.

Not all materials are crystalline, although for most solids the crystalline state is the natural one because the energy of the ordered atomic arrangement is usually lower than that of an irregular packing of atoms. However, when the atoms are not given the opportunity to arrange themselves in an orderly manner, by inhibiting their mobility during solidification, an *amorphous* material may be formed. This is the case, for example, in the formation of soot. In other cases, the molecules may be extremely long and irregular in shape, so that an orderly arrangement may not be obtained easily, as in the case of polymers. In some materials, such as glass, the solid state corresponds to a *supercooled liquid* in which the molecular arrangement of the liquid state is frozen in. Due to the high viscosity of the liquid, crystals do not have time to grow under normal conditions, and an amorphous material is formed. Upon annealing, such glassy materials may crystallize (*devitrify*), as observed in the case of quartz.

In the remaining chapters we shall have ample opportunity to indicate the importance of the regular or irregular stacking of atoms on the properties of materials.

## References

R. Beeching, *Electron Diffraction*, 2nd ed., Methuen, London, 1946.

C. W. Bunn, *Chemical Crystallography*, Oxford, New York, 1945.

R. W. James, *X-Ray Crystallography*, 4th ed., Methuen, London, 1950.

L. Pauling, *Nature of the Chemical Bond*, Cornell University Press, Ithaca, 1945.

## Problems

**1.1** Given that one gram molecule of a gas at 0°C and a pressure of 760 mm mercury occupies a volume of 22.414 liters, and assuming Avogadro's number is $6.025 \times 10^{23}$, compute the number of molecules per $m^3$ in a gas at 0°C and 760 mm mercury (Loschmidt's number).

**1.2** A residual pressure of $10^{-10}$ mm mercury in a vacuum tube is considered very good vacuum; estimate the number of gas molecules per $m^3$ in such a tube at room temperature.

**1.3** According to the kinetic theory of gases, the average kinetic energy of a gas molecule at an absolute temperature $T$ is equal to $(3/2)kT$, where $k$ is Boltzmann's constant. What is the average energy, expressed in electron volts, at room temperature ($T = 300°K$)? If the gas is hydrogen, what is the order of magnitude of the velocity of the molecules at $T = 300°K$?

**1.4** Calculate the velocity of an electron with a kinetic energy of 1 ev; what is the velocity of a proton with a kinetic energy of 1 ev?

**1.5** Calculate the kinetic energy, the potential energy, and the total energy of an electron in the ground state of a hydrogen atom according to the theory of Bohr.

**1.6** Calculate the energy and radii of the first four Bohr orbits for an electron in a hydrogen atom.

**1.7** An electron in a hydrogen atom makes a transition from a quantum state of principal quantum number $n = 2$ to the ground state. What is the energy and what is the frequency $f$ of the emitted light quantum? In what region of the electromagnetic spectrum do you place this frequency?

**1.8** According to wave mechanics, the wavelength $\lambda$ of an electron is related to the momentum $p$ of the electron by means of the so-called de Broglie formula $\lambda = h/p$, where $h$ is Planck's constant. Show that the wavelength of an electron with kinetic energy of $V$ electron volts is given by $\lambda = (150/V)^{1/2}$ angstroms.

**1.9** Show that Bohr's quantum postulate for circular orbits is equivalent to the statement that the circumference of the orbit is equal to an integer times the wavelength of the electron.

**1.10** In an electron-diffraction experiment one wishes to have a wavelength of the electrons of 0.5 angstrom. What accelerating voltage is required to obtain this wavelength?

**1.11** According to wave mechanics, the charge distribution corresponding to the electron in the ground state of a hydrogen atom is an exponential function of the type $\rho(r) = A \exp(-2r/r_1)$ where $A$ and $r_1$ are constants and $r$ represents the distance from the nucleus. Given that the total charge must be equal to $-e$, show that $A = -e/\pi r_1^3$ (see formula 1.11).

**1.12** Show that the integrand in formula (1.12) has its maximum value for $r = r_1$.

**1.13** On the basis of the rules pertaining to the possible values of the orbital and magnetic quantum numbers $l$ and $m_l$, set up a table which gives all possible quantum states for the principal quantum number $n = 3$ [in analogy with (1.13) for $n = 2$].

**1.14** Assume the energy of two particles in the field of each other is given by the following function of the distance $r$ between the centers of the particles:

$$W(r) = -(\alpha/r) + (\beta/r^8)$$

where $\alpha$ and $\beta$ are constants.

(a) Show that the two particles form a stable compound for $r = r_0 = (8\beta/\alpha)^{1/7}$.

(b) Show that in the stable configuration the energy of attraction is 8 times the energy of repulsion (in contrast with the fact that the attractive force is equal to the repulsive force!).

(c) Show that the total potential energy of the two particles in the stable configuration is equal to

$$-(7/8)(\alpha^8/8\beta)^{1/7} = -(7/8)\alpha/r_0.$$

(d) Show that if the particles are pulled apart, the molecule will break as soon as $r = (36\beta/\alpha)^{1/7} = r_0(4.5)^{1/7}$, and that the minimum force required to break the molecule is equal to

$$[\alpha^{9/7}/(36\beta)^{2/7}][1 - 8/(36)^{2/7}].$$

**1.15** Suppose an atom A has an ionization energy of 5 ev, and an atom B has an electron affinity of 4 ev (i.e., an energy of 4 ev is gained by attaching a free electron to atom B). Suppose atoms A and B are 5 angstroms apart. What is the energy required to transfer an electron from A to B?

**1.16** The edge of the elementary cube of a body-centered cubic lattice is $a$ meter. How many atoms are there on the average per cube of $a^3$ meter$^3$? Answer the same questions for a face-centered cubic lattice and for a simple cubic lattice (atoms only at the corners of the elementary cube).

**1.17** Suppose identical atoms are arranged in a simple cubic lattice; the atoms may be considered as hard spheres of radius $R$. The edge of the elementary cube, $a$, is equal to $2R$ so that neighboring atoms touch each other. Show that the fraction of the volume occupied by atoms is $\pi/6 = 0.523$.

**1.18** Consider a body-centered cubic lattice of identical atoms; the atoms may be considered as hard spheres of radius $R$. Atoms along a body diagonal touch each other. Show that the fraction of the volume occupied by atoms in this arrangement is $(\pi\sqrt{3})/8 = 0.68$.

**1.19** Consider a face-centered cubic lattice of identical atoms with radius $R$. Atoms along a face diagonal touch each other. Show that the fraction of the volume occupied by atoms in this arrangement is $(\pi\sqrt{2})/6 = 0.74$. (Note that this is the most economical way of stacking identical spheres; compare answers to problems 1.17 and 1.18.)

**1.20** Two kinds of atoms, A and B, form a crystal with the same structure as CsCl. Considering the atoms as hard spheres of radii $r_a$ and $r_b$, show that the atoms along a body diagonal of the elementary cube cannot touch each other if the ratio $r_a/r_b$ (or $r_b/r_a$) is larger than 1.37.

**1.21** Two elements A and B form a compound AB which crystallizes in the sodium chloride structure. Assuming the atoms may be considered as hard spheres of radii $r_a$ and $r_b$, show that atoms along a cube edge cannot touch each other as soon as the ratio of the radii is larger than 2.44.

**1.22** Copper crystallizes in a face-centered cubic lattice, the cube edge being 3.608 angstroms at room temperature. A single crystal of copper has been cut so that the surface of the crystal is parallel to one of the faces of the elementary cube. A monochromatic beam of X-rays with a wavelength of 1.658 angstroms is incident on the surface of the crystal. Show that the planes parallel to the surface reflect the X-rays if the angle between the beam and the surface is approximately 27 deg or 67 deg.

**1.23** The X-ray beam with a wavelength of 1.658 angstroms mentioned in the preceding problem is obtained by electron bombardment of a nickel target. What is the minimum anode voltage required in the X-ray tube to produce this wavelength? Suppose that instead of X-rays one uses electrons of the same wavelength, what would be the required accelerating voltage? (see problem 1.8).

**1.24** From the data given for copper in problem 1.22, calculate the number of atoms per $m^3$ in this material.

# 2

# *Dielectric Properties of Insulators in Static Fields*

In this chapter we deal with the behavior of insulators in static fields; their behavior in alternating fields will be discussed in the next chapter. It would undoubtedly be more elegant to start immediately with time-dependent fields, and to treat the static behavior as a particular case corresponding to zero frequency. From the student's point of view, however, there are advantages in dealing with the simpler case first because it gives him a chance to absorb the fundamental concepts at a more leisurely pace. Once these concepts are well understood, the transition to the concept of the complex dielectric constant and its interpretation becomes a good deal easier.

The questions to be discussed in this chapter are of the following nature: What is the relationship between the *macroscopic* measurable dielectric constant and the *atomic structure* of a material? Why do some materials have a high and others a low dielectric constant? Why do the dielectric constants of some materials depend on temperature, whereas in other cases they do not? What happens to the dielectric constant when a substance melts or solidifies? These and other questions will be discussed in terms of simplified atomic "models." These models are not necessarily the best representation of atoms known to physicists, but they serve mainly to illustrate the basic ideas underlying more sophisticated calculations. The first two sections are devoted to the macroscopic theory of the dielectric constant as one finds it expounded in much more detail in textbooks on field theory. The remainder of the chapter deals with the atomic interpretation of the dielectric constant.

## 2.1 The static dielectric constant

The reader may be reminded at this point of some fundamentals concerning electric fields. One of the most useful theorems in this area is that of Gauss. It states that the total electric flux $\phi$ emanating from a closed surface is equal to the total charge enclosed by that surface. Denoting the charges enclosed by the surface by $Q_1, Q_2, \ldots, Q_i, \ldots, Q_n$, where the $Q$'s may be positive and negative, this theorem may be expressed mathematically by means of a surface integral as follows:

$$\phi = \iint \mathbf{D} \cdot d\mathbf{S} = \sum_{i=1}^{n} Q_i \tag{2.1}$$

Here, $\mathbf{D}$ represents the *flux density* in coulombs m$^{-2}$ at the center of the surface element represented by the outwardly directed vector $d\mathbf{S}$; the integration extends over the entire closed surface. When one deals with a continuous charge distribution of density $\rho$ instead of with discrete charges, the sum on the right-hand side of (2.1) must, of course, be replaced by the volume integral of $\rho$, the integration extending over the entire volume enclosed by the surface under consideration.

The *electric field strength* $\mathbf{E}$ in any point of space, i.e. the force per unit charge, is related to the flux density in that point by

$$\mathbf{D} = \epsilon_0\epsilon_r\mathbf{E} \tag{2.2}$$

Here, $\epsilon_0 = 8.854 \times 10^{-12}$ farad m$^{-1}$ represents the dielectric constant or permittivity of a vacuum; $\epsilon_r$ is called the *relative dielectric constant* or the *relative permittivity* of the material. It is important to emphasize that $\epsilon_0$ enters only as a result of using a particular system of units, in our case mks units; $\epsilon_0$ therefore has no other physical meaning than that of a fundamental conversion factor. The relative dielectric constant, $\epsilon_r$, on the other hand is determined by the atomic structure of the material and it is with the physical interpretation of this quantity that we shall deal below. Note that $\epsilon_r$ is a dimensionless quantity which is equal to 1 for vacuum. For all substances, $\epsilon_r > 1$ for reasons that will be explained. It should also be mentioned here that expression (2.2) refers only to *isotropic materials;* i.e., to materials for which the dielectric and other physical properties are independent of the direction in which they are measured. In crystals, for example, the dielectric constant generally depends on the direction along which it is measured relative to the crystal axes. In polycrystalline materials, on the other hand, with a random distribution of the grains, the directional effects disappear and (2.2) is applicable. For single crystals one cannot, in general, use (2.2) and the dielectric constant should then be replaced by a tensor quantity. Unless stated otherwise, we shall assume isotropic materials. From the fact that $E$ is expressed in newton coulomb$^{-1}$ and from the fact that $\epsilon_r$ is dimensionless, the reader can readily verify from (2.2) that $\epsilon_0$ has the dimensions of farad m$^{-1}$.

A method for measuring $\epsilon_r$ for a particular material emerges from the following reasoning: consider a parallel plate condenser as indicated in Fig. 2.1. The area of the plates is $A$ and the distance between them $d$;

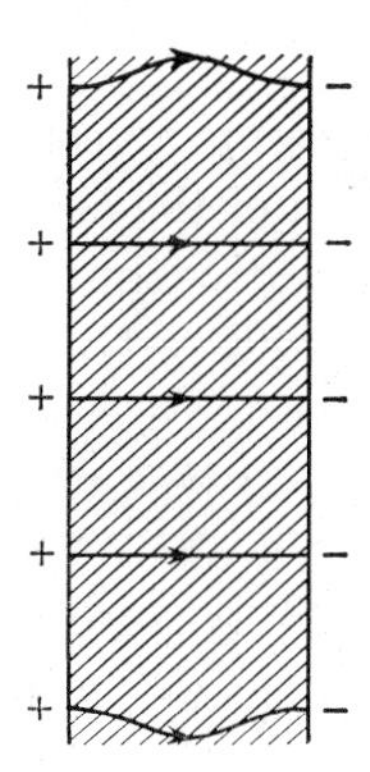


**Fig. 2.1.** Charged parallel plate condenser; flux lines are indicated.

assume the charge per unit area on the plates is $+q$. Neglecting end-effects, the flux lines run from the positive to the negative plate in a direction perpendicular to the plates. By applying the theorem of Gauss in a suitable manner the reader will readily verify that the magnitude of the flux density is given by $D = q$; this will be true whether or not the space between the plates is filled with an insulating material. It thus follows from (2.2) that the field strength in the region between the plates is given by

$$E = D/\epsilon_0\epsilon_r = q/\epsilon_0\epsilon_r \tag{2.3}$$

Since the voltage difference between the two plates is simply given by $Ed$ (homogeneous field!), the capacitance of the system is equal to $C = qA/Ed = \epsilon_0\epsilon_r A/d$. Hence, if $C_{\text{vac}}$ represents the capacitance when the space between the plates is evacuated, one immediately finds $\epsilon_r$ from the relation

$$\epsilon_r = C/C_{\text{vac}} \tag{2.4}$$

Thus, $\epsilon_r$ can be determined experimentally be measuring the capacitance with and without the dielectric.

## 2.2 Polarization and dielectric constant

In this section we shall show that in a dielectric subjected to an electric field $E$, each volume element may be thought of as carrying an *electric dipole moment* which is proportional to the field strength. As we shall see later, the result obtained is of fundamental importance because it provides a link between the macroscopic dielectric constant and the atomic theory of this quantity.

The electric dipole moment of a neutral system of point charges $Q_1, Q_2, \ldots, Q_i, \ldots, Q_n$ is defined as a vector given by

$$\boldsymbol{\mu} = \sum_i Q_i \mathbf{r}_i \tag{2.5}$$

where $\mathbf{r}_i$ represents a vector drawn from the origin of a coordinate system to the position of the charge $Q_i$; the charges $Q_i$ may, of course, be either positive or negative, but their sum $\Sigma\, Q_i$ must be zero to comply with the neutrality of the system. As shown in problem 2.8 at the end of this chapter, the vector $\boldsymbol{\mu}$ is independent of the choice of the origin of the coordinate system; if it were not, it would be a useless concept. In mks units, an electric dipole moment is evidently expressed in coulomb meters. In its simplest form, a dipole moment consists of two equal point charges of opposite sign, $\pm Q$, separated by a distance $d$. Choosing the origin of the coordinate system to coincide with the negative charge, the dipole moment in this case has a magnitude equal to $Qd$, and is represented by a vector pointing from the negative charge in the direction of the positive charge, as indicated in Fig. 2.2.

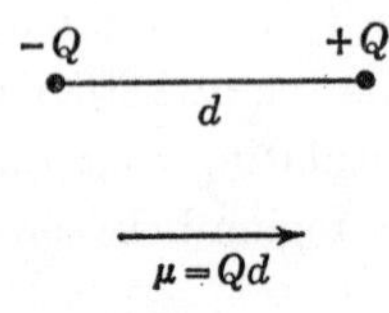


**Fig. 2.2.** Dipole and dipole moment vector $\mu$ for two equal charges of opposite sign separated by a distance $d$.

Let us now turn to the statement made in the beginning of this section by considering the simple case of a homogeneous and isotropic dielectric subjected to a homogeneous electric field $E$ produced between two charged parallel plates. Let the flux density be $D = \epsilon_0\epsilon_r E$. Suppose now that we were to cut out of the dielectric a small volume element $dx\ dy\ dz$ where $dx$ is chosen perpendicular to the plates, as indicated in Fig. 2.3. This would evidently result in a distortion of the field; i.e., it would no longer be homogeneous. Let us consider the following question: how can the field be kept homogeneous, and equal to the original field $E$? This can be achieved in two ways:

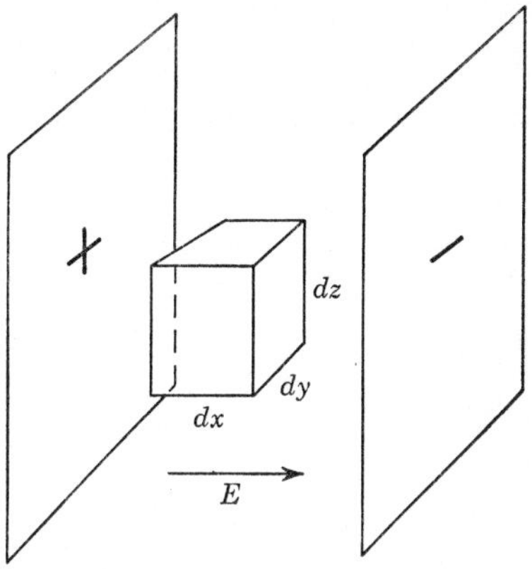

**Fig. 2.3.** Illustrating a cavity $dx\ dy\ dz$ cut away from a dielectric between two charged plates.

(a) By inserting the material again in the cavity produced. This answer is trivial, but nevertheless plays a role in the argument, as we shall see later.

(b) In attempting to produce a homogeneous field in the presence of the cavity, we evidently require that if $E_i$ and $E_o$ represent the field strengths respectively inside and outside the cavity

$$E_i = E_o = E$$

Since inside the cavity there is no material, this requirement, when expressed in terms of the flux densities $D_i$ and $D_o$ inside and outside the cavity, becomes

$$D_i/\epsilon_0 = D_o/\epsilon_0\epsilon_r = D/\epsilon_0\epsilon_r \tag{2.6}$$

Hence, if the flux density inside the cavity is made $\epsilon_r$ times as small as the flux density outside the cavity, we have succeeded in making the field homogeneous in the presence of the cavity. Now, according to the theorem of Gauss, a change in flux density at a surface can be achieved only if the surface carries an electric charge. Thus, with reference to Fig. 2.4, if

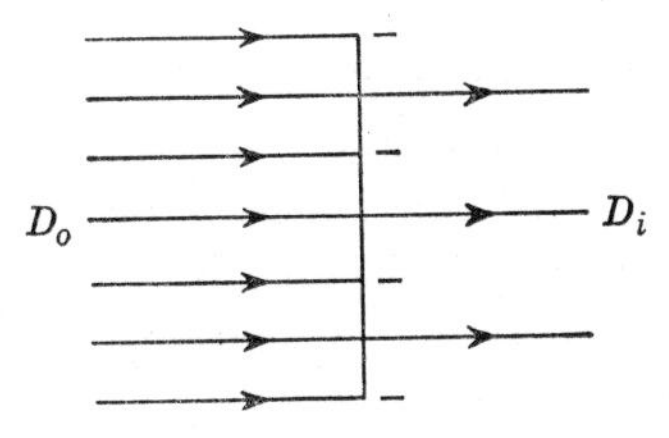


**Fig. 2.4.** Illustrating reduction of a flux density $D_o$ in passing across a plane to the value $D_i$ as a result of the presence of negative charges on the surface.

we desire a flux density $D_o$ to be reduced to a flux density $D_i$ upon crossing a surface, the surface should be provided with a charge density of $-(D_o - D_i)$ coulomb m$^{-2}$. In our problem, therefore, the field can be made homogeneous by placing a negative charge of $-(D_o - D_i)\ dy\ dz$ on the left-hand face of the cavity in Fig. 2.3, and a positive charge of $(D_o - D_i)\,dy\,dz$ on the right-hand face. This system of charges is apparently neutral, and corresponds to a dipole moment

$$(D_o - D_i)\,dx\,dy\,dz \tag{2.7}$$

The direction of the dipole moment vector is from left to right in Fig. 2.3, i.e. parallel to the applied field **E**. Now, according to (2.6) we have $D_o/D_i = \epsilon_r$, so that (2.7) may be written as

$$D_i(\epsilon_r - 1)\,dx\,dy\,dz = \epsilon_0(\epsilon_r - 1)E\,dx\,dy\,dz \tag{2.8}$$

What conclusion can we draw from these arguments? Since answer (a) achieves the same result as providing the cavity with a dipole moment given by (2.8) we conclude that the material which previously occupied the cavity carried a dipole moment given by (2.8). Thus, *a dielectric subjected to a homogeneous field carries a dipole moment* ***P*** *per unit volume* which, according to (2.8) may be written as

$$\mathbf{P} = \epsilon_0(\epsilon_r - 1)\mathbf{E} \tag{2.9}$$

The dipole moment per unit volume **P** is called the *polarization* of the dielectric. It is expressed in coulomb $m^{-2}$ and is proportional to the field strength as long as $\epsilon_r$ is independent of $E$, which it is for normal dielectrics below the breakdown field. It is emphasized that in the derivation of (2.9) nothing was said about the physical state of the dielectric; hence, it is valid for gases, liquids and solids. It will become evident below that (2.9) provides the *link between the macroscopic and atomic theory of dielectrics.*

## 2.3 The atomic interpretation of the dielectric constant of monoatomic gases

The simplest, though for most purposes not the most practical, kinds of materials are the rare gases such as helium and argon. These gases are simple from a theoretical point of view because, first of all, in a gas the average distance between the atoms or molecules is large enough so that one can neglect interaction between them, and furthermore, if we restrict ourselves to the rare gases, the molecules consist of single atoms. Forgetting for the moment the discussion of the preceding two sections, let us consider the problem from the atomic point of view and investigate what can be said about the properties of a rare gas when it is subjected to an electric field $E$. First consider a single atom consisting of a positive nucleus of charge $Ze$, and $Z$ electrons moving around the nucleus. Since the nucleus has a diameter of the order of $10^{-15}$ m, whereas the radius of the electron cloud is of the order of $10^{-10}$ m, we may consider the nucleus for our purpose as a point charge. As a crude model for the electron cloud, let us assume that the total negative charge $-Ze$ is distributed homogeneously throughout a sphere of radius $R$, where $R \approx 10^{-10}$ m

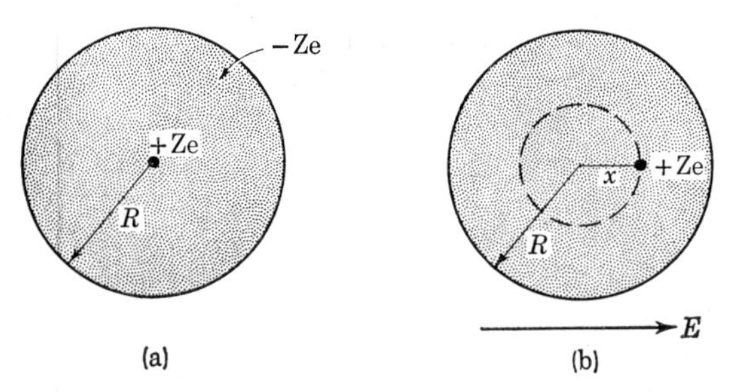


**Fig. 2.5.** Atomic model in the absence of a field is given in (a); the shift of the negative charge cloud relative to the nucleus resulting from the field $E$ is presented schematically in (b). In practice, $x \ll R$.

[see Fig. 2.5(a)]. Although this model is a far cry from what one knows about atoms, the results that we obtain give the correct order of magnitude for the quantities of interest. When this atomic model is placed in the field **E**, the nucleus and the electric cloud will evidently try to move in opposite directions because of the opposite signs of their charges. However, as they are pulled apart, a force will develop between them which tends to drive the nucleus back to the center of the sphere. Consequently, an equilibrium will be obtained in which the nucleus is shifted slightly relative to the center of the electron cloud in the direction of **E**. Quantitatively, this shift may be calculated for this model as follows: assume that in equilibrium with the field **E**, the nucleus is displaced by the amount $x$ as indicated in Fig. 2.5(b); we shall assume here that the shape of the electron cloud is not influenced by the field; i.e., it is assumed to remain a sphere of radius $R$. The force on the nucleus along the field direction is $ZeE$. The electron cloud can be divided into two regions: one inside an imaginary sphere of radius $x$, and one between the two spherical surfaces of radii $x$ and $R$. By applying Gauss theorem, the reader will readily verify that the charge in the latter region does not exert a force on the nucleus. The only force exerted on the nucleus is that produced by the negative charge inside the sphere of radius $x$; the charge inside this sphere is equal to $-Zex^3/R^3$. The force exerted by this charge on the nucleus can be obtained by concentrating the charge in the center and applying

Coulomb's law; since the total force on the nucleus must be zero in equilibrium, we obtain

$$ZeE = \frac{1}{4\pi\epsilon_0}\frac{(Ze)(Zex^3/R^3)}{x^2} \tag{2.10}$$

Hence, in equilibrium the nucleus will be displaced relative to the center of the sphere by the amount

$$x = (4\pi\epsilon_0 R^3/Ze)E \tag{2.11}$$

Note that $x$ is proportional to the field strength and that the problem dealt with is analogous to that in which a mechanical force is exerted on a particle bound with an elastic force to a certain equilibrium position.

What have we learned by deriving (2.11)? In order to see this, consider first the atom in the absence of the field $E$. If we were to probe the space outside the atom with an infinitesimally small test charge, we would detect no field at all, because the system is neutral and has no dipole moment (nucleus coincides with the center of the negative charge cloud!). In the presence of the field, the system is still neutral, but has a non-zero *dipole moment* because the nucleus and the center of the charge cloud are separated by a distance $x$. The atom will thus appear to carry a dipole moment (see section 2.2) equal to

$$\mu_{\text{ind}} = Zex = 4\pi\epsilon_0 R^3 E = \alpha_e E \tag{2.12}$$

Here, the subscript "ind" refers to the word "induced"; the dipole moment is induced by the field because it was not there in the absence of the field. The *induced dipole moment* is proportional to the field strength and the proportionality factor $\alpha_e$ is called the *electronic polarizability* of the atom; "electronic," because the dipole moment results from a shift of the electron cloud relative to the nucleus. Note that $\alpha_e$ is proportional to $R^3$, i.e. to the volume of the electron cloud.

So far, we have considered only a single atom. Consider now a rare gas containing $N$ atoms per m$^3$, subjected to a homogeneous field $E$. Neglecting any interaction between the dipoles induced in the atoms, which is a good approximation for a gas, we find for the polarization of the gas, i.e. the electric dipole moment per unit volume

$$\mathbf{P} = N\alpha_e\mathbf{E} \tag{2.13}$$

Comparing this expression with the macroscopic equation (2.9) for $\mathbf{P}$, we conclude that for rare gases

$$\epsilon_0(\epsilon_r - 1) = N\alpha_e \tag{2.14}$$

In other words, we have obtained a *relationship between the measurable quantity $\epsilon_r$ and the atomic constant $\alpha_e$*. Let us now investigate to what extent there exists agreement between theory and experiment. The dielectric constant of He, measured at 0°C and 1 atmosphere, is found experimentally to be $\epsilon_r = 1.0000684$; under these conditions, the gas contains approximately $N = 2.7 \times 10^{25}$ atoms per $m^3$. For the model used above we find from (2.12) and (2.14)

$$\epsilon_r - 1 = 4\pi NR^3 \tag{2.15}$$

Calculating $R$ from this expression and the numerical values of $\epsilon_r$ and $N$, one obtains $R \cong 0.6 \times 10^{-10}$ meter, which is indeed the correct order of magnitude for the radius of an atom. Thus, even though the model is rather crude, the results indicate that the interpretation is essentially correct. By way of illustration we give here the polarizability $\alpha_e$ of rare gas atoms in $10^{-40}$ farad $m^2$ as units.

| | He | Ne | A | Kr | Xe |
|---|---|---|---|---|---|
| $\alpha_e$ . . . | 0.18 | 0.35 | 1.43 | 2.18 | 3.54 |

Note that the polarizability increases as the atoms become larger, in agreement with the results obtained for the model.

Let us now estimate the order of magnitude of the relative shift $x$ of the nucleus and the center of the electron cloud, because this is the quantity which, together with the number of atoms per unit volume, determines $\epsilon_r$. For a field of, say, $10^5$ volts per meter we obtain from (2.11) with $R \approx 10^{-10}$ m and $Z \approx 10$, $x \cong 7 \times 10^{-18}$ m, which is very small indeed compared to the radius of the atom. The perturbing influence of an applied field on an atom is apparently very slight.

We note that the electronic polarizability of an atom is determined completely by its electronic structure; as long as the structure remains the same, $\alpha_e$ remains the same. This notion is important because the electron structure of an atom is essentially independent of temperature, unless the temperature is extremely high. Thus, for normal temperatures, $\alpha_e$ is *independent of temperature*. Consequently, if the number of atoms per unit volume is kept constant, the dielectric constant will also be independent of temperature. This is indeed what has been found experimentally for the rare gases. If the number of atoms per unit volume is allowed to change, the dielectric constant is simply proportional to this number.

## 2.4 Qualitative remarks on the dielectric constant of polyatomic molecules

The dielectric constant of polyatomic gases depends in many cases on temperature even if the number of molecules per $m^3$ is kept constant; in other cases it may be constant. Some examples are given in Fig. 2.6 and in this section we shall deal with the physical interpretation of these ob-

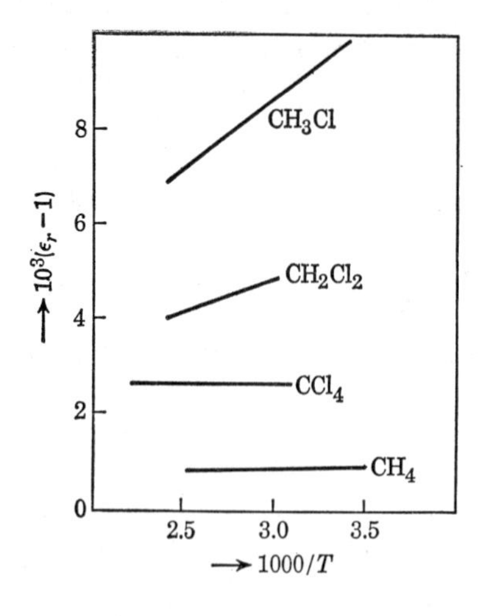


**Fig. 2.6.** The relative dielectric constant as a function of temperature (°K) for some gases at a pressure of 1 atmosphere.

servations. First of all, it should be realized that if two different atoms A and B form a chemical bond, one of the two is more apt to part with one or more of its valence electrons than the other. Thus, when $Z_{\mathrm{A}}e$ and $Z_{\mathrm{B}}e$ represent the nuclear charges of the two atoms, and if atom A has the tendency to give valence electrons to B, one finds on the average more than $Z_{\mathrm{B}}$ electrons around the nucleus B and fewer than $Z_{\mathrm{A}}$ electrons around the nucleus A. One says that atom A is more *electropositive* than B. Consequently, the bond between A and B is at least partly ionic (it is not necessary that A parts with an integral number of electrons; an electron may spend more time near B than near A, so that on the average, A has parted only with a fraction of the electron). If the bond between A and B has ionic character, it is evident that the molecule AB carries an *electric dipole moment even in the absence of an applied field.* For obvious reasons, such a dipole moment is called *permanent*, and will be denoted by the vector $\boldsymbol{\mu}_p$. The magnitude of the dipole moment is given by the product of the average charge transferred from A to B and the internuclear distance. For a molecule consisting of more than two atoms, several bonds may carry a permanent dipole moment and the resulting permanent dipole moment of the molecule as a whole is obtained by vector addition of the moments associated with the various bonds. This is illustrated in Fig. 2.7 for two hypothetical cases corresponding to the type ABA. It is observed that the resultant dipole moment may be zero, viz. if the molecule has a center of symmetry.

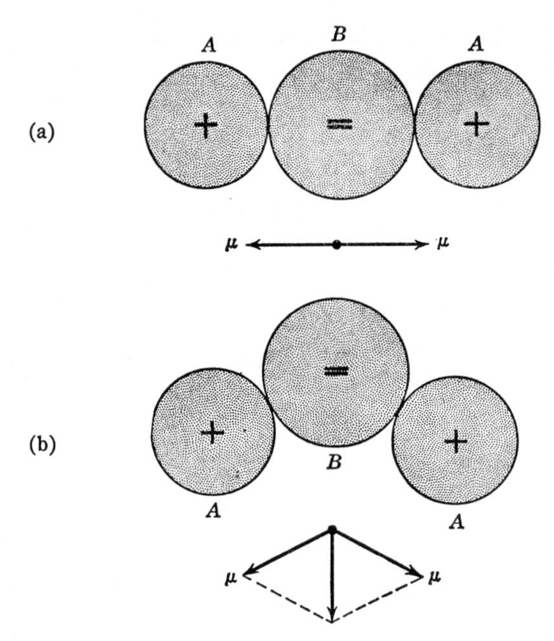


**Fig. 2.7.** Two possible configurations of a molecule $A_2B$. In (a) the resultant dipole moment is zero; in (b) there is a resultant dipole moment given by the vector sum of the dipole moments of the individual ionic bonds.

It will be evident that molecules such as $O_2$, $H_2$, $N_2$ etc., which consist of similar atoms, carry no permanent dipole moment.

When an external field $\mathbf{E}$ is applied to a molecule carrying a permanent dipole moment $\boldsymbol{\mu}_p$, the external field will tend to align $\boldsymbol{\mu}_p$ along the direction of $\mathbf{E}$, since $\mathbf{E}$ exerts a torque on $\boldsymbol{\mu}_p$. The contribution of this process of orientation of the permanent dipoles to the polarization $\mathbf{P}$ is called the *orientational polarization*, and will be denoted by $\mathbf{P}_o$. Furthermore, the effect of an external field will be to shift the electron clouds in the molecule relative to the respective nuclei. In a molecular gas, therefore, we also have electronic polarization, as in a monoatomic gas, the only difference being that the molecules consist of more than one atom. In analogy with the electronic polarizability of a single atom, one defines the electronic polarizability of a molecule as the dipole moment induced per unit field strength resulting only from shifts of the electron clouds relative to the nuclei, i.e.

$$\boldsymbol{\mu}_{\text{ind}} = \alpha_e \mathbf{E} \tag{2.16}$$

It is understood that this induced dipole moment represents an average value, the average being taken over all possible orientations of the molecule relative to the field; this stipulation is necessary because the induced dipole moment of a molecule such as AB evidently depends on the angle between the direction of **E** and the line joining the nuclei of the molecule.

Besides the two contributions to the polarization of a molecular gas just mentioned, there is a third contribution which is referred to as the *ionic polarization*. This contribution takes account of the fact that when in a molecule some of the atoms have an excess positive or negative charge (resulting from the ionic character of the bonds), an electric field will tend to shift positive ions relative to negative ones. This leads to an induced moment of different origin from the moment induced by electron clouds shifting relative to nuclei. The difference between ionic and electronic polarization may be illustrated with reference to the string of ions represented in Fig. 2.8. The electronic polarizability measures the shift of

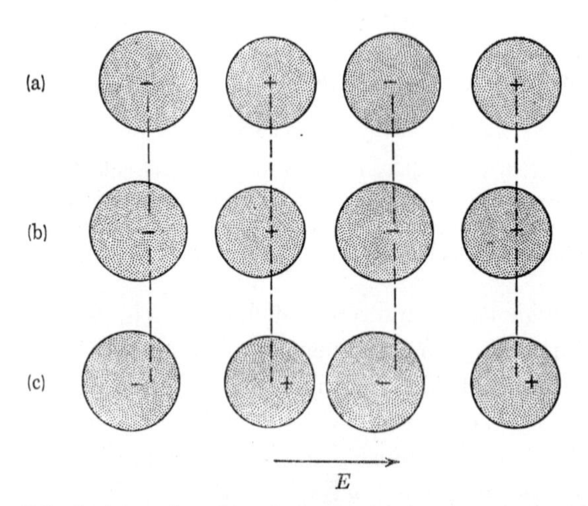


**Fig. 2.8.** A string of positive and negative ions is represented in (a) in the absence of a field. In (b) all electron clouds are shifted to the left as a result of the field, but the nuclei have been fixed; this corresponds to electronic polarization. In (c) the polarized ions are displaced relative to each other; the positive ions to the right, the negative ions to the left. In (c) therefore there is electronic as well as ionic polarization.

the electron clouds in the ions relative to the nuclei to which they belong. The ionic polarizability measures the shift of the ions relative to each other. A molecule may thus be characterized by an ionic polarizability $\alpha_i$ defined in a way similar to $\alpha_e$.

Summarizing this section, we can distinguish between the following contributions to the polarization in a polyatomic gas:

(i) the *orientational polarization* ($P_o$);
(ii) the *electronic polarization* ($P_e$);
(iii) the *ionic polarization* ($P_i$).

The total polarization is given by the sum of these three quantities.

## 2.5 Quantitative discussion of the dielectric constant of polyatomic gases

Consider a gas containing $N$ molecules per m$^3$. We shall assume the molecules carry a permanent electric dipole moment $\boldsymbol{\mu}_p$; the electronic and ionic polarizabilities of the molecules will be denoted respectively by $\alpha_e$ and $\alpha_i$. The questions we shall answer in this section are these: What is the relationship between the dielectric constant $\epsilon_r$ and the atomic quantities given? Does this relationship agree with experimental information of particular gases, and if so, what can one learn from such information?

From what has been said in the preceding section, we can write down the electronic and ionic contributions to the total polarization immediately as

$$P_e + P_i = N(\alpha_e + \alpha_i)E \tag{2.17}$$

where $E$ is the applied field. We mentioned before that $\alpha_e$ is independent of the temperature $T$ as long as the electronic structure of the molecule remains unaltered. By the same token, $\alpha_i$ will be independent of temperature if the electronic structure does not change. We thus conclude that for the usual temperatures of interest, $(\alpha_e + \alpha_i)$ may be considered independent of $T$.

A calculation of the orientational polarization $P_o$ is somewhat more complicated; we shall see that $P_o$ does depend on temperature. Consider a system of $N$ permanent dipoles of magnitude $\mu_p$ in the absence of a field; let the temperature be $T$. Since there is no preference for any particular direction, the vector sum of all the individual $\boldsymbol{\mu}_p$'s will vanish; i.e., there is no polarization in the absence of a field. Suppose we choose an arbitrary direction and call this the $x$-axis. The number of molecules $N(\theta)\, d\theta$ for which the direction of $\boldsymbol{\mu}_p$ at a given instant lies within an angle between $\theta$ and $(\theta + d\theta)$ with the $x$-axis is then simply proportional to the solid angle $2\pi \sin\theta\, d\theta$, as indicated in Fig. 2.9. Suppose now, a field $\mathbf{E}$ is applied along the $x$-direction. The number of molecules for which the direction of $\boldsymbol{\mu}_p$ at a given instant now lies within an angle between $\theta$ and $(\theta + d\theta)$ with the $x$-axis is no longer proportional to $2\pi \sin\theta\, d\theta$, for the

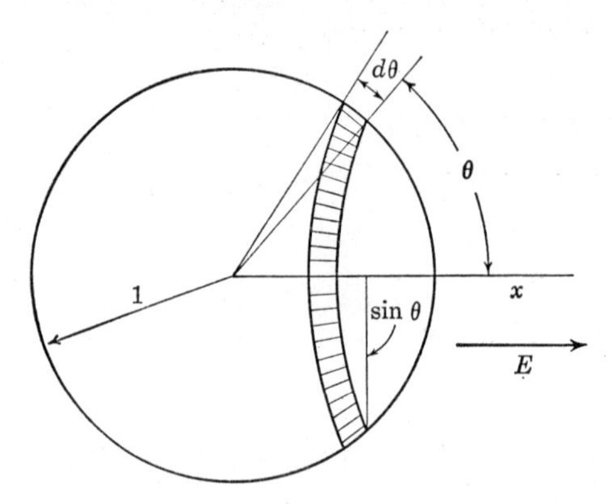


**Fig. 2.9.** Illustrating the geometry used in calculating $N(\theta)\,d\theta$. The area of the shaded ring between $\theta$ and $\theta + d\theta$ is equal to $2\pi \sin\theta\, d\theta$.

simple reason that the field **E** makes the $x$-axis a preferred direction; we shall show in fact, that the number is given by

$$N(\theta)\,d\theta = A2\pi \sin\theta\, d\theta \exp\left[(\mu_p E \cos\theta)/kT\right] \qquad (2.18)$$

where $k$ is Boltzmann's constant ($= 1.38 \times 10^{-23}$ joule per degree C), and $A$ is a constant of proportionality determined by the total number of dipoles $N$ under consideration. To derive (2.18), first consider the energy of a dipole $\boldsymbol{\mu}_p$ in the external field **E**. As indicated in Fig. 2.10, the torque

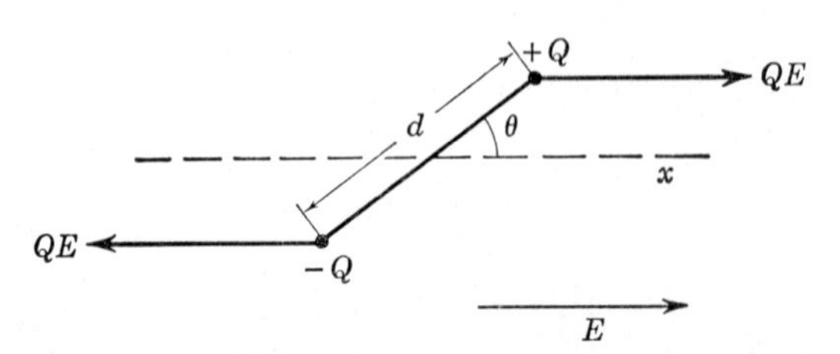


**Fig. 2.10.** Illustrating the forces exerted by a field $E$ on the two charges of a dipole. The torque produced is equal to $QEd \sin\theta = \mu_p E \sin\theta$.

produced by the field on $\boldsymbol{\mu}_p$ is equal to $\mu_p E \sin\theta$. Let us arbitrarily set the energy of a dipole in the field equal to zero if $\theta = 90$ deg. The energy $W(\theta)$ of the dipole for an arbitrary angle $\theta$ is then given by

$$W(\theta) = \int_{\theta=90^\circ}^{\theta} \mu_p E \sin\theta\, d\theta = -\mu_p E \cos\theta = -\boldsymbol{\mu}_p \cdot \mathbf{E} \qquad (2.19)$$

From Fig. 2.11 we see that the energy is lowest for a dipole $\boldsymbol{\mu}_p$ parallel to **E**, and highest for a dipole antiparallel to **E**. In other words, *small angles $\theta$ are preferred over large ones.* In fact, if there were no thermal motion,

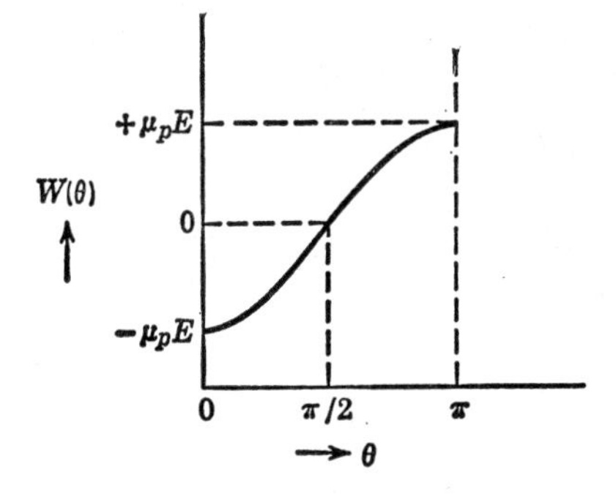


**Fig. 2.11.** The potential energy of a dipole $\mu_p$ as a function of the angle $\theta$ between the dipole vector and an applied field.

all dipoles would line up along the external field direction. The reader who is familiar with the elements of Boltzmann statistics will now recognize that the population corresponding to a solid angle $2\pi \sin\theta \, d\theta$ must be weighted by a Boltzmann factor $\exp[-W(\theta)/kT]$ so that (2.18) follows immediately from this and (2.19).

Our next task is to find an expression for $P_o$ on the basis of our result (2.18). It will be evident that $\mathbf{P}_o$ is parallel to $\mathbf{E}$; i.e., we are interested in finding the sum of the *components* of the individual $\boldsymbol{\mu}_p$ along the direction of $\mathbf{E}$. Since a dipole $\boldsymbol{\mu}_p$ which makes an angle $\theta$ with $\mathbf{E}$ has a component along $\mathbf{E}$ equal to $\mu_p \cos\theta$, one may write

$$P_o = \int_{\theta=0}^{\pi} N(\theta) \, d\theta \, \mu_p \cos\theta \tag{2.20}$$

Substituting (2.18) into (2.20) and making use of the fact that $N = \int_0^\pi N(\theta) \, d\theta$, the constant $2\pi A$ may be eliminated and one obtains

$$P_o = N \frac{\int_0^\pi \mu_p \cos\theta \exp[(\mu_p E \cos\theta)/kT] \sin\theta \, d\theta}{\int_0^\pi \exp[(\mu_p E \cos\theta)/kT] \sin\theta \, d\theta} \tag{2.21}$$

Introducing the new variable $y = (\mu_p E \cos\theta)/kT$ and writing for convenience $\mu_p E/kT = a$, the reader may verify that this leads to

$$P_o = NkT \frac{\int_{-a}^{+a} ye^y \, dy}{\int_{-a}^{+a} e^y \, dy} = N\mu_p \left[\coth a - \frac{1}{a}\right] \equiv N\mu_p L(a) \tag{2.22}$$

The function $L(a)$ defined by (2.22) is called the *Langevin function;* it first appeared in a study by Langevin (1905) of the similar problem of orientation of magnetic dipoles in a magnetic field. The function $L(a)$ is represented in Fig. 2.12. For large values of $a$ it approaches unity and hence $P_o$ approaches $N\mu_p$. Physically, this corresponds to the situation in which $\mu_p E/kT$ is so large that one approaches complete alignment of

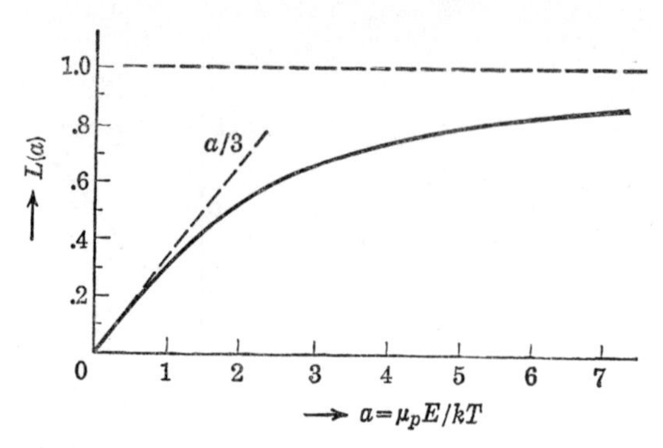


**Fig. 2.12.** The Langevin function $L(a)$; for $a \ll 1$, $L(a) \cong a/3$.

the dipoles along the field direction. In practice, this approach to *saturation* of the orientational polarization is never encountered in the case of gases; in fact, for practical purposes we may assume $a = \mu_p E/kT \ll 1$. This can be seen as follows: the magnitude of a permanent dipole moment may be anticipated to be of the order of the product of an electronic charge and one angstrom. It is for this reason that electric dipole moments are usually expressed in so-called debye units:

$$1 \text{ debye unit} = 10^{-10} \text{ esu angstrom} = 3.33 \times 10^{-30} \text{ coulomb meter}$$

Debye postulated the existence of permanent dipoles in molecules in 1912 and has contributed a great deal to our present understanding of dielectrics. Assuming for the moment permanent dipole moments of 1 debye unit, one finds even for a strong field of $10^7$ volts per meter at room temperature, $a \approx 0.01$, which is small compared to 1. Under these circumstances expression (2.22) simplifies to

$$P_o = N\mu_p^2 E/3kT \tag{2.23}$$

Thus, the *orientational polarization is inversely proportional to the temperature and proportional to the square of the permanent dipole moment.* Note that the slope of the Langevin function for small values of $a$ is $\frac{1}{3}$. Also note that this derivation shows clearly that the tendency of the external field to align the dipoles is counteracted by the thermal motion, resulting in a decreasing value of $P_o$ with increasing $T$.

The total polarization of a polyatomic gas is given by the sum of (2.17) and (2.23), i.e.

$$P = N(\alpha_e + \alpha_i + \mu_p^2/3kT)E \tag{2.24}$$

Comparing this with the general relation (2.9) from the *macroscopic* theory, we find that the dielectric constant $\epsilon_r$ is related to the molecular properties as follows:

$$\epsilon_0(\epsilon_r - 1) = N(\alpha_e + \alpha_i + \mu_p^2/3kT) \tag{2.25}$$

Let us now turn to the question of how this result can be compared with experimental information. It is observed that if the dielectric constant $\epsilon_r$ is plotted as a function of $1/T$, expression (2.25) predicts the result to be a straight line. The slope of this line is determined by $\mu_p^2$; the intercept with the axis $1/T = 0$ provides a measure for $\alpha_e + \alpha_i$. In Fig. 2.6 we showed some experimental results which indeed confirm the temperature dependence predicted by (2.25). From such measurements one can evidently calculate the permanent dipole moment of the molecules as well

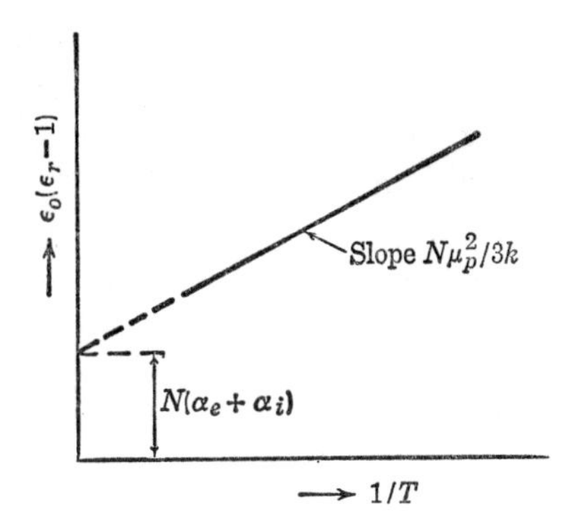


**Fig. 2.13.** Illustrating the relationship between $\epsilon_r$ and $1/T$ as predicted by expression (2.25). From the slope one can determine the permanent dipole moment of the molecules, and from the intercept of the ordinate, obtained by extrapolation, the sum $(\alpha_e + \alpha_i)$ may be found.

as the sum $\alpha_e + \alpha_i$, provided the number of molecules per m³ is known, as indicated in Fig. 2.13. By way of illustration, we give in Table 2.1 some permanent dipole moment in debye units obtained from measurements of the dielectric constant in the gas phase. From the results obtained,

**Table 2.1.** Experimentally determined permanent dipole moments of various molecules (in debye units, $3.33 \times 10^{-30}$ coulomb meter)

| Molecule | $\mu_p$ | Molecule | $\mu_p$ |
|---|---|---|---|
| NO | 0.1 | $CO_2$ | 0 |
| CO | 0.11 | $CS_2$ | 0 |
| HCl | 1.04 | $H_2O$ | 1.84 |
| HBr | 0.79 | $H_2S$ | 0.93 |
| HI | 0.38 | $CH_4$ | 0 |
| $NO_2$ | 0.4 | $CH_3Cl$ | 1.15 |

one may derive certain conclusions with regard to the structure of the molecules. For example, the fact that $CO_2$ has no resultant dipole moment, whereas each of the CO bonds does have a dipole moment, indicates that in this molecule the two bonds make an angle of 180 degrees with each other; the $CO_2$ molecule thus must look like this: O=C=O. On the other hand, a molecule such as $H_2O$ does have a resultant $\boldsymbol{\mu}_p$, which indicates that the two OH bonds make an angle different from 180 degrees, and that the molecule has a triangular form (see Fig. 2.7). Dielectric constant measurements have been used a great deal as a tool for investigating molecular structure, but this subject lies outside the scope of the present text.

## 2.6 The internal field in solids and liquids

A detailed interpretation of the dielectric properties of solids and liquids is considerably more complicated than for gases, but a semi-quantitative understanding may be achieved on the basis of the concepts developed in the preceding sections. The main problem which arises in the case of solids and liquids is the calculation of what is known as the *local* or *internal field* $\mathbf{E}_i$, which is defined as the field acting at the location of a given atom. In the case of a gas, we assumed that the internal field was equal to the applied field $\mathbf{E}$, and as long as the density of molecules is reasonably low, this is a good approximation. However, in solids and liquids the molecules or atoms are so close together that the field seen by a given particle is determined in part by the dipoles carried by surrounding particles; in general, therefore, $\mathbf{E}_i$ *is not equal to the applied field* $\mathbf{E}$.

To illustrate this for a simple example, consider an infinite string of similar equidistant atoms of polarizability $\alpha_e$, as indicated in Fig. 2.14.*

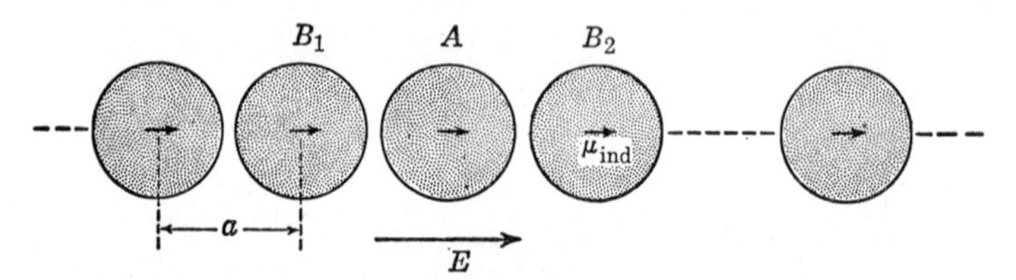


**Fig. 2.14.** Illustrating a string of atoms of polarizability $\alpha_e$ in an external field parallel to the string. The induced dipole moments are indicated.

Given an external field $\mathbf{E}$ applied in a direction parallel to the string, what is the internal field at the position of a given atom? With reference to Fig. 2.14 let the field seen by atom A be $\mathbf{E}_i$; from the symmetry of the problem it is evident that $\mathbf{E}_i$ will be parallel to $\mathbf{E}$ and furthermore, in the

* The author noticed this particular model in the present context for the first time in a set of lecture notes of Professor D. J. Epstein (M.I.T.), who kindly gave his permission to include it in this book.

example chosen, the field seen by the other atoms will be the same. The dipole moment induced in each of the atoms of the string is thus

$$\boldsymbol{\mu}_{\text{ind}} = \alpha_e \mathbf{E}_i \tag{2.26}$$

and our problem is to evaluate $\mathbf{E}_i$. Clearly, $\mathbf{E}_i$ must be equal to the applied field $\mathbf{E}$ plus the field produced at the location of A by the dipoles on all other atoms. Let us proceed to calculate the field produced at the center of A by the dipole on atom $B_1$ in Fig. 2.14; we shall assume that the dipoles may be considered as point dipoles. According to field theory, the potential around a point dipole $\boldsymbol{\mu}$ in vacuum is given by

$$V(r, \theta) = \frac{1}{4\pi\epsilon_0}\frac{\mu \cos\theta}{r^2} \tag{2.27}$$

where $r$ is the distance from the dipole and $\theta$ is the angle between $\mathbf{r}$ and $\boldsymbol{\mu}$. The field around a dipole, therefore, has two components given by

$$E_r = -\frac{\partial V}{\partial r} = \frac{1}{4\pi\epsilon_0}\frac{2\mu\cos\theta}{r^3} \tag{2.28}$$

and

$$E_\theta = -\frac{1}{r}\frac{\partial V}{\partial \theta} = \frac{1}{4\pi\epsilon_0}\frac{\mu \sin\theta}{r^3} \tag{2.29}$$

These components are indicated in Fig. 2.15. Thus, the field produced by $B_1$ at the location of atom A is obtained from the last two expressions by

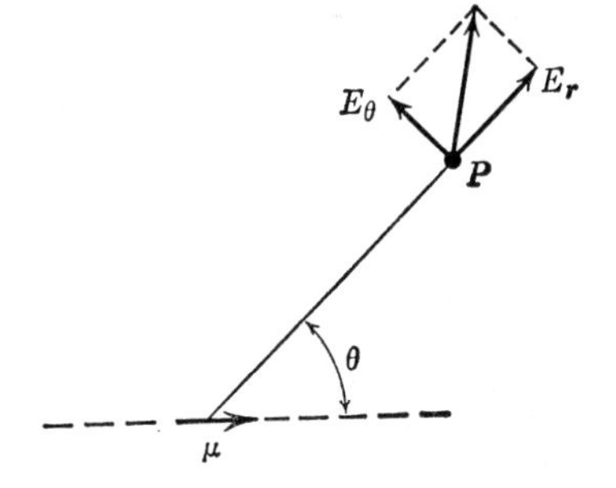


**Fig. 2.15.** Illustrating the field components $E_r$ and $E_\theta$ in a point $P$ resulting from a dipole $\boldsymbol{\mu}$.

putting $r = a$ and $\theta = 0$; this gives a contribution of $\mu_{\text{ind}}/2\pi\epsilon_0 a^3$ in the direction of $\mathbf{E}$. It is readily verified that the field produced at A by the dipole on $B_2$ is equal to that produced by $B_1$. Following the same procedure for the other atoms in the string, we find for the internal field at A

$$E_i = E + \left(\frac{\mu_{\text{ind}}}{\pi\epsilon_0 a^3}\right)\sum_{n=1}^{\infty}\frac{1}{n^3} \tag{2.30}$$

where $n$ accepts the integer values 1, 2, 3 . . . . Substituting $\mu_{\text{ind}}$ from

(2.26) into (2.30) we can express $E_i$ in terms of the applied field $E$. The sum in (2.30) is approximately equal to 1.2, and we obtain

$$E_i = \frac{E}{1 - 1.2\alpha_e/\pi\epsilon_0 a^3} = \frac{E}{1 - \beta} \tag{2.31}$$

Since the constant $\beta$ is positive, we conclude that the *actual field seen by an atom in the string is larger than the applied field E.* Physically this means that for the model chosen here, the dipoles *cooperate* with each other in the sense that a large dipole moment on a given atom helps induce a dipole moment in its neighbors, which in turn induce a dipole moment in the former, etc. It is observed that the *cooperation becomes stronger as the polarizability of the atoms increases and as the distance between them decreases.* It is also observed that the internal field is determined in general by the *structure* of a given material, i.e. by the surroundings seen at the position of a given atom. An accurate calculation of the internal field in solids and liquids is in general very complicated and for our purposes it may be sufficient to point out a general feature of the relation between the internal and applied fields. We note that the last term in (2.30) is proportional to the dipole moment induced in the atoms. It is not surprising therefore that calculations of $E_i$ in a three-dimensional case always give results which may be written in the form

$$E_i = E + (\gamma/\epsilon_0)P \tag{2.32}$$

where $P$ is the dipole moment per unit volume and where $\gamma$ is a proportionality constant which is referred to as the *internal field constant;* $\epsilon_0$ has been introduced in (2.32) only for the purpose of making $\gamma$ a dimensionless quantity. In general, the numerical value of $\gamma$ is of the order of unity; for the linear chain of Fig. 2.14, for example, it follows immediately from (2.30) that $\gamma = 1.2/\pi$. A particular case which one encounters frequently in the literature on dielectrics is that corresponding to $\gamma = \frac{1}{3}$. One speaks in that case of the *Lorentz field,* given by

$$E_{i\,\text{Lorentz}} = E + P/3\epsilon_0 \tag{2.33}$$

A derivation of this expression may be found in the references given at the end of this chapter; it holds only in particular cases, as when the atoms in a solid are surrounded cubically by other atoms. One should, therefore, be careful in applying the specific value $\gamma = \frac{1}{3}$, unless the particular symmetry conditions are met for which it has been derived. A further warning should be given here pertaining to the interpretation of the dielectric constant of liquids in which the molecules carry permanent dipoles (*polar liquids*). The internal field as it tends to orient the dipoles along the direction of the external field is generally *different* from the internal field as it is used to calculate the contribution to the polarization resulting from $\alpha_e$ and $\alpha_i$. At first sight, this may seem strange; in fact, this difference was not realized until it was pointed out by Onsager in 1936. In the material before that time, this oversight had resulted in erroneous interpretations of the dielectric behavior of polar liquids. For a discussion of the internal field in liquids we refer the reader to Böttcher's book.

## 2.7 The static dielectric constant of solids

In this section the general features of the dielectric behavior of solids will be discussed; for data concerning specific materials the reader is referred to *Dielectric Materials and Applications*, edited by von Hippel. The special behavior of ferroelectric and piezoelectric materials is dealt with in subsequent sections.

It is convenient to distinguish between three groups of solids in connection with their dielectric behavior:

(i) **Elemental dielectrics.** These are materials built up from only one kind of atoms, such as diamond, phosphorus, etc. It will be evident that in such materials there are no permanent dipoles or ions, so that the only contribution to the polarization is that due to the relative displacement of electron clouds and nuclei. Hence, for these materials $P = P_e$ and $P_i = P_o = 0$. Denoting the electronic polarizability per atom by $\alpha_e$, we may write

$$P = N\alpha_e E_i \tag{2.34}$$

where $E_i$ is the internal field and $N$ represents the number of atoms per m³. In writing expression (2.34) it has been assumed that the internal field is the same for all atoms. Since $E_i$ is generally given by an expression like (2.32), we may write

$$P = N\alpha_e[E + (\gamma/\epsilon_0)P] \tag{2.35}$$

from which we can find $P$ in terms of the applied field $E$ as

$$P = \frac{N\alpha_e E}{1 - (\gamma N\alpha_e/\epsilon_0)} \tag{2.36}$$

In order to obtain an expression for the dielectric constant $\epsilon_r$ in terms of the atomic quantities we turn to the macroscopic expression (2.9); from it and (2.36) we then find

$$\epsilon_0(\epsilon_r - 1) = \frac{N\alpha_e}{1 - \gamma N\alpha_e/\epsilon_0} \tag{2.37}$$

In case we assume $\gamma = \frac{1}{3}$ (see preceding section), the same procedure leads to the so-called Clausius-Mosotti expression

$$\frac{\epsilon_r - 1}{\epsilon_r + 2} = \frac{N\alpha_e}{3\epsilon_0} \tag{2.38}$$

The dielectric constant is thus determined by $N$, $\alpha_e$ and $\gamma$. It should be noted that in general, $\alpha_e$ is not the same as the polarizability of the free atoms, because the binding between the atoms affects the valence electrons; however, the two values may be nearly the same. The distance between the atoms in a solid is affected only slightly by temperature, and therefore $N$, $\alpha_e$, $\gamma$, and the dielectric constant $\epsilon_r$ are in first approximation independent of temperature for the materials under discussion. By way of illustration we give here the dielectric constant $\epsilon_r$ for three elements in the fourth group of the periodic table, all three having the diamond structure.

| | Diamond | Si | Ge |
|---|---|---|---|
| $\epsilon_r$ . . . | 5.68 | 12 | 16 |

Although silicon and germanium are poor insulators, one can still speak of a dielectric constant for such materials. It may be worthwhile to point out that $(\epsilon_r - 1)$ for these and for most other solids is of the order of unity or ten, whereas for gases at normal temperature and pressure the same quantity is of the order of $10^{-3}$ or $10^{-4}$. This difference of course reflects the difference in the number of atoms per unit volume, as may be seen from (2.37). In fact, a representative figure for the number of atoms per m³ in a solid or liquid is $N \cong 5 \times 10^{28}$ m$^{-3}$. Accurate values in specific cases may be obtained from X-ray diffraction data; such data provide information about the crystal structure and the interatomic distances.

(ii) **Ionic dielectrics without permanent dipoles.** In ionic crystals such as the alkalide halides, the total polarization is made up of electronic and ionic polarization, i.e.

$$P = P_e + P_i \tag{2.39}$$

A crystal of this kind, when considered as a huge molecule, has no permanent electric dipole moment, because the sum

$$\boldsymbol{\mu}_p = \sum_i e_i \mathbf{r}_i$$

(see section 2.2) vanishes; hence, $P_o = 0$. The dielectric behavior of such materials is more complicated than that of group (i) from the point of view of quantitative interpretation; the internal field at the positive ion sites, for example, is in general different from that at the negative ion sites.

Without going into details with reference to the interpretation, it may be of interest to point out that the ionic polarization $P_i$ usually constitutes a considerable fraction of the total polarization. Experimentally this has been established as follows: when one measures the dielectric constant at frequencies of approximately $5 \times 10^{14}$ sec$^{-1}$, corresponding to the visible part of the electromagnetic spectrum, the relatively heavy positive and negative ions can no longer follow the rapid field variations. Consequently, one measures in that region only the electronic polarization $P_e$. Since a measurement of the static dielectric constant gives $P_e + P_i$, it is thus possible to find $P_e$ and $P_i$ separately. Quantitatively this may be expressed in the following manner: Let $\epsilon_{rs}$ represent the static value of the relative dielectric constant; we may then write in accordance with (2.9)

$$\epsilon_0(\epsilon_{rs} - 1)E = P_e + P_i \tag{2.40}$$

Similarly, when $\epsilon_{re}$ represents the dielectric constant measured at optical frequencies, one may write

$$\epsilon_0(\epsilon_{re} - 1)E = P_e \tag{2.41}$$

From these expressions it is obvious that the *difference between* $\epsilon_{rs}$ *and* $\epsilon_{re}$ *is a measure for the ionic polarization.* The values of $\epsilon_{rs}$ and $\epsilon_{re}$ given in Table 2.2 for alkali halides illustrate clearly that the static dielectric constant of these materials contains an appreciable contribution from the displacement of the positive ion lattice relative to the negative ion lattice in an external field. It should be mentioned here that in the optical region of the electromagnetic spectrum, one measures the dielectric constant by determining the *index of refraction n.* According to Maxwell's theory of electromagnetic waves, $\epsilon_{re} = n^2$ for materials with a magnetic permeability equal to that of vacuum.

**Table 2.2.** Static and optical dielectric constants, $\epsilon_{rs}$ and $\epsilon_{re}$, for alkali halides; $n$ represents the index of refraction

| Solid | $\epsilon_{rs}$ | $\epsilon_{re} = n^2$ | Solid | $\epsilon_{rs}$ | $\epsilon_{re} = n^2$ |
|---|---|---|---|---|---|
| LiF | 9.27 | 1.92 | KF | 6.05 | 1.85 |
| LiCl | 11.05 | 2.75 | KCl | 4.68 | 2.13 |
| LiBr | 12.1 | 3.16 | KBr | 4.78 | 2.33 |
| LiI | 11.03 | 3.80 | KI | 4.94 | 2.69 |
| NaF | 6.0 | 1.74 | RbF | 5.91 | 1.93 |
| NaCl | 5.62 | 2.25 | RbCl | 5.0 | 2.19 |
| NaBr | 5.99 | 2.62 | RbBr | 5.0 | 2.33 |
| NaI | 6.60 | 2.91 | RbI | 5.0 | 2.63 |

It may be of interest to point out that there exists a relationship between the difference ($\epsilon_{rs} - \epsilon_{re}$) and the *compressibility* of these materials. This may be seen as follows: The compressibility of a material is defined as the fractional change in volume per unit change in pressure. It is evident that the compressibility of a material will be small for "hard" atoms and large for "soft" atoms. In the case of an ionic solid, the applied field tends to shift the positive ion lattice in a direction opposite to that in which the negative ions tend to move, so that for a system of "hard" ions, the ionic polarization per unit applied field will be smaller than for a system of "soft" ions. Thus, in general one expects large values of ($\epsilon_{rs} - \epsilon_{re}$) to be accompanied by large values of the mechanical compressibility of the material; this notion is in agreement with experimental evidence.

(iii) **Solids containing permanent dipole moments.** The molecules in many solids carry permanent electric dipole moments; solid nitrobenzene ($C_6H_5NO_2$) is an example of this group. The dielectric constant in the vicinity of the melting point of this material is represented in Fig. 2.16.

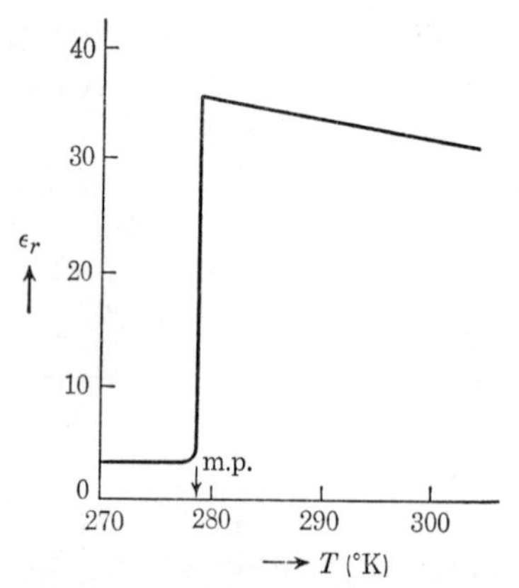


**Fig. 2.16.** Relative dielectric constant of nitrobenzene as a function of temperature in the vicinity of the melting point. [After C. P. Smyth and C. S. Hitchcock, *J. Am. Chem. Soc.*, **55**, 1296 (1933)]

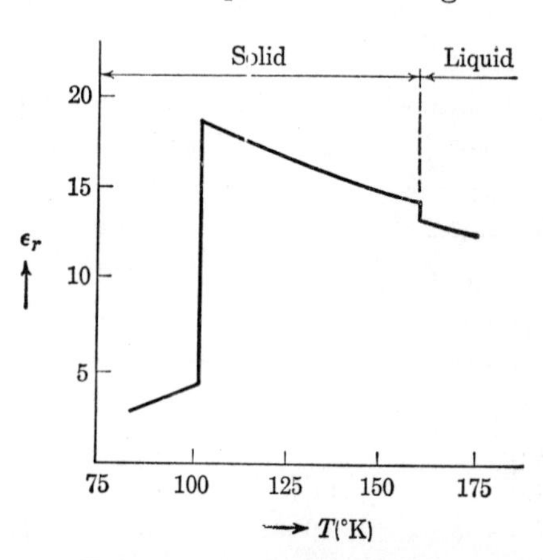


**Fig. 2.17.** The relative dielectric constant of solid and liquid HCl as a function of temperature. [After C. P. Smyth and C. S. Hitchcock, *J. Am. Chem. Soc.*, **55**, 1830 (1933)]

It is observed that as the material freezes, the dielectric constant $\epsilon_r$ decreases abruptly from a value near 35 to a much lower value. It is also observed that in the solid state $\epsilon_r$ is independent of temperature, whereas in the liquid state it decreases with increasing temperature. These observations can be understood qualitatively on the basis of the concepts developed in the preceding sections. In the liquid state, $\epsilon_r$ is determined by the electronic, ionic, and orientational contributions. The decrease of $\epsilon_r$ with increasing temperature must evidently be ascribed to the fact that the orientational contribution decreases with temperature, as it does in the case of gases [see formula (2.25)]. The sudden drop in $\epsilon_r$ at the melting point is interpreted as meaning that even though the solid contains permanent dipoles, these are "frozen" in the solid state and can no longer be aligned by an external field. Thus, in the solid state one only measures the contributions to $\epsilon_r$ associated with electronic and ionic polarization. This also explains the fact that in the solid state the temperature dependence of $\epsilon_r$ has disappeared.

In some solids, the permanent dipoles may still contribute to the polarization, although their motion may be strongly inhibited. An example is given in Fig. 2.17. It is observed that as HCl goes from the liquid state to the solid state at 159°K, $\epsilon_r$ increases abruptly by a small amount; this is due to the change in density upon solidification. However, in the solid state, $\epsilon_r$ keeps increasing with decreasing temperature, indicating the presence of orientational polarization. It is not until a temperature of 100°K has been reached that $\epsilon_r$ drops sharply, a result of the fact that the dipoles have become immobile.

After what has been said concerning the static dielectric constant of solids, there is little to be added about the same quantity in the case of liquids. In general, $\epsilon_r$ will decrease with increasing temperature as a result of a reduction in the orientational polarization. If there are no permanent dipoles present, $\epsilon_r$ is nearly independent of temperature.

## 2.8 Some properties of ferroelectric materials

For the dielectric materials discussed in the preceding sections, the polarization is a linear function of the applied field. There are, however, a number of substances for which the polarization of a specimen depends on its history; i.e., the polarization in these materials is not a unique function of the field strength. In particular, these materials exhibit *hysteresis* effects, similar to those observed in ferromagnetic materials; they are therefore called *ferroelectric* materials. An example of a hysteresis loop associated with the polarization versus field strength is given in Fig. 2.18. When an electric field is applied to a "virgin" specimen of a ferroelectric material, the polarization increases along a curve such as $OABC$ in Fig. 2.18. When the field is reduced, it is observed that for $E = 0$, a certain amount of *remanent polarization*, $P_r$, is still present. In other

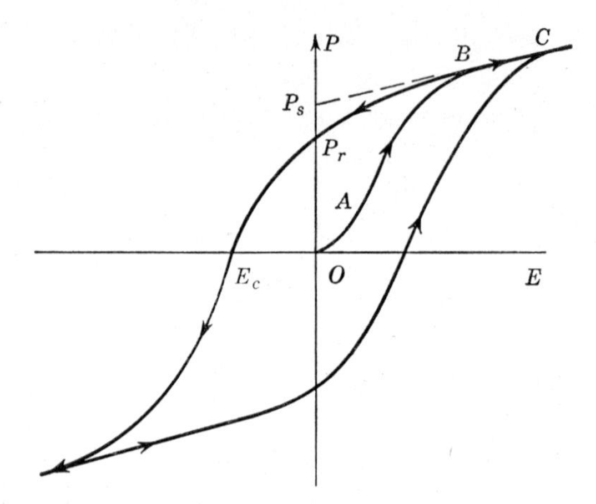


**Fig. 2.18.** Schematic representation of a hysteresis curve for a ferroelectric material. $P_r$ represents the remanent polarization, $P_s$ the spontaneous magnetization. The slope along $BC$ is due to "normal" dielectric polarization. $E_c$ is the coercive field.

words, the material is *spontaneously polarized*. In order to make the polarization equal to zero, a field in the opposite direction must be applied; this field is called the *coercive field*, and is denoted by $E_c$ in Fig. 2.18.

The hysteresis loop may be explained qualitatively in the following manner: the direction of the spontaneous polarization is generally not the same throughout a macroscopic specimen. In fact, the specimen may be considered to consist of a number of *domains* which are themselves spontaneously polarized, but with the direction of polarization varying from one domain to another. Thus, a virgin macroscopic specimen may have zero polarization as a whole; i.e., the resultant of the polarization vectors of the individual domains may vanish. Upon application of an electric field, the domains for which the polarization points along the direction of the applied field grow at the expense of other domains for which the polarization points in other directions. This process corresponds to the curve $OAB$ in Fig. 2.18. Ultimately, the specimen may have become one single domain, and the further slight increase of $P$ with increasing applied field is due to "normal" polarization as discussed in preceding sections. The domain structure can be studied in ferroelectric materials, for example, by employing polarized light, which makes domains visible. In Fig. 2.19 we have illustrated schematically how the polarization of a crystal of $BaTiO_3$ (barium titanate) may change direction under the influence of a field which has a direction opposite to that of the spontaneous polarization. Domains with a polarization parallel to the applied field evidently

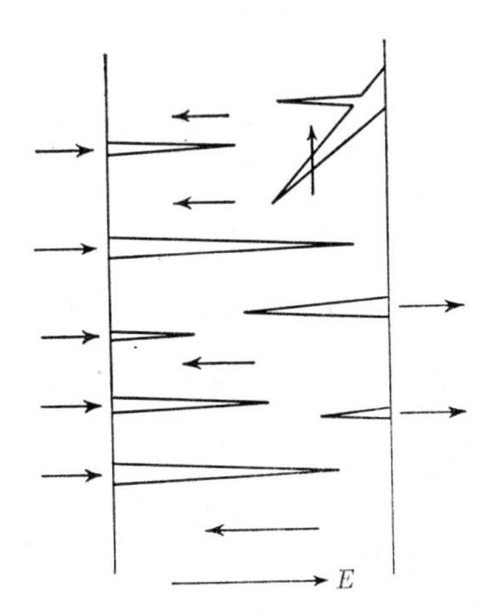


**Fig. 2.19.** Schematic representation of new domains resulting from application of a field $E$ directed oppositely to the spontaneous polarization of a specimen.

grow in the form of thin needles of approximately $10^{-6}$ m width; these domains grow essentially in the *forward* direction until ultimately the polarization of the specimen lies along the applied field direction. This process is quite different from that encountered in ferromagnetism, as we shall see in Chapter 4.

The *spontaneous polarization*, which is the most characteristic property of a ferroelectric material, usually vanishes above a certain temperature $\theta_f$; this temperature is called the ferroelectric *Curie temperature.* In the ferroelectric region, i.e. below $\theta_f$, the dielectric constant is evidently a function of the field strength and is no longer a "constant." One can, of course, define a differential relative dielectric constant defined on the basis of (2.9) by the equation

$$\epsilon_0(\epsilon_r - 1) = dP/dE \tag{2.42}$$

When one speaks of "the dielectric constant" in the ferroelectric region one usually means $\epsilon_r$ defined by (2.42) along the virgin curve at the origin. The dielectric constant so defined may reach very high values in the vicinity of the ferroelectric Curie temperature, as may be seen from Fig. 2.20 for barium titanate ceramic.

Above the Curie temperature, the dielectric constant varies with temperature in accordance with the so-called Curie-Weiss law

$$\epsilon = C/(T - \theta) \tag{2.43}$$

where $C$ is a constant and $\theta$ is a characteristic temperature which is usually a few degrees smaller than the ferroelectric Curie temperature $\theta_f$.

*Classification of ferroelectric materials.* There are various groups of ferroelectric materials which may be classified on the basis of their chemical composition and structure.

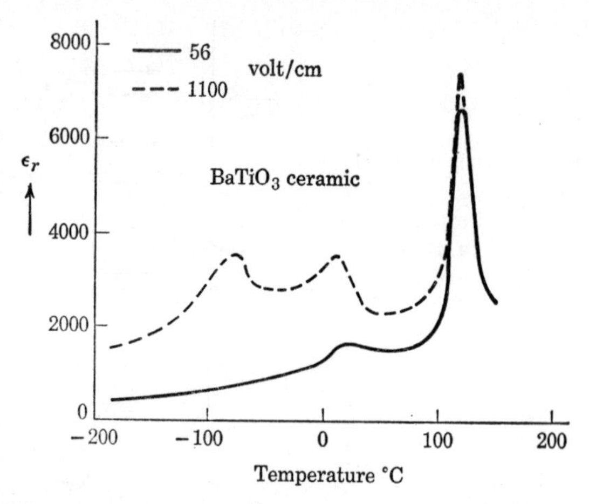


**Fig. 2.20.** Dielectric constant of barium titanate ceramic as a function of temperature. The fully drawn and the dashed curves correspond respectively to a peak field strength at 1 kc of 56 and 1100 volts per cm. The sharp peaks occur at the ferroelectric Curie temperature $\theta_f$. (After W. B. Westphal, Laboratory for Insulation Research, M.I.T.)

(i) The first solid which was recognized to exhibit ferroelectric properties is *Rochelle salt*, the sodium-potassium salt of tartaric acid ($NaKC_4H_4O_6 \cdot 4H_2O$). It has the peculiar property of being ferroelectric only in the temperature region between $-18$°C and 23°C; i.e., it has two transition temperatures. The spontaneous polarization of this material

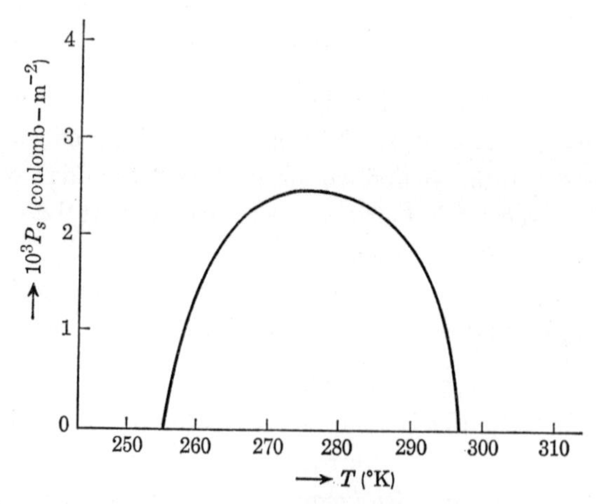


**Fig. 2.21.** Spontaneous polarization of Rochelle salt as a function of temperature. [After J. Halblützel, *Helv. Phys. Acta*, **12**, 489 (1939)]

is represented in Fig. 2.21. Rochelle salt is representative of the *tartrate group* of ferroelectric materials; other members of this group are those in which a fraction of the potassium in Rochelle salt is replaced by $NH_4$, Rb or Tl.

(ii) In 1935 Busch and Scherrer discovered ferroelectric properties in $KH_2PO_4$, which is a typical example of the *dihydrogen phosphates* and *arsenates* of the alkali metals. The spontaneous polarization of this material is given in Fig. 2.22 as a function of temperature. The shape of this curve resembles that of the spontaneous magnetization in iron, as we shall see in Chapter 4. In this case, there is only one Curie temperature, viz. $\theta_f = 123°K$.

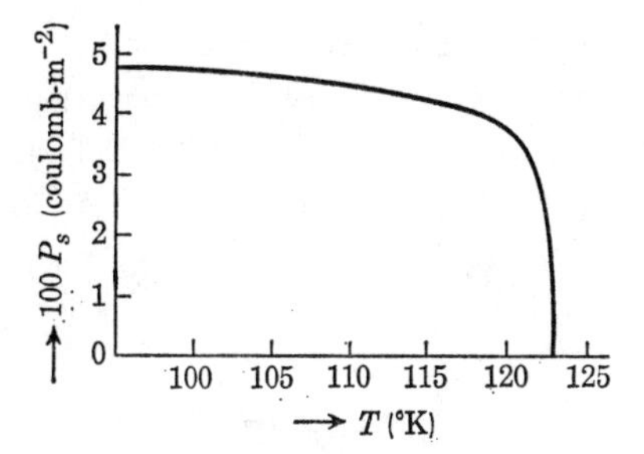


**Fig. 2.22.** Spontaneous polarization of $KH_2PO_4$ as a function of temperature. [After A. von Arx and W. Bantle, *Helv. Phys. Acta*, **16**, 221 (1943)]

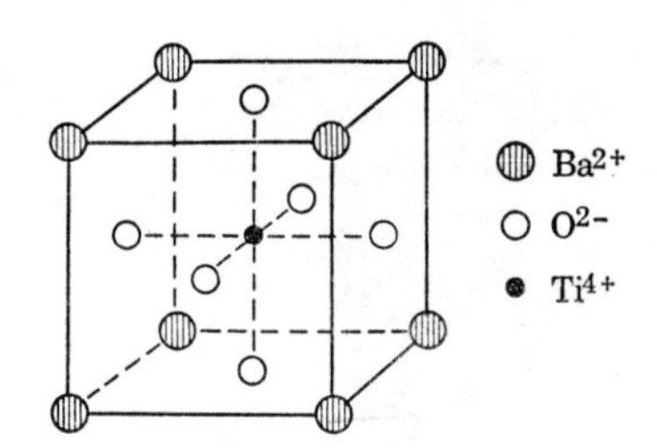


**Fig. 2.23.** The cubic structure of $BaTiO_3$ above the ferroelectric Curie temperature.

(iii) Probably the best-known ferroelectric material is *barium titanate* $BaTiO_3$; it is a representative of the so-called *oxygen octahedron group* of ferroelectric materials. The reason for this name is that above the Curie temperature ($\theta = 120°C$), $BaTiO_3$ corresponds to the cubic structure presented in Fig. 2.23. In this structure, the $Ba^{2+}$ ions occupy the corners of a cube; the centers of the cube faces are occupied by $O^{2-}$ ions. The oxygen ions form an octahedron, at the center of which the small $Ti^{4+}$ ion is located. The $Ti^{4+}$ ion is considerably smaller than the space which is available inside the oxygen octahedron. It thus brings with it a *high ionic polarizability* for two reasons: (a) it has a charge of $4e$ and, (b) it can be displaced over a relatively large distance. We shall see that this may be the explanation for the occurrence of spontaneous polarization in $BaTiO_3$.

There is an intimate relationship between the ferroelectric properties and the atomic arrangement in ferroelectric materials. Above 120°C, $BaTiO_3$ has the cubic structure indicated in Fig. 2.23. When the temperature is lowered through the critical temperature of 120°C, the material becomes spontaneously polarized and at the same time the structure changes. The direction of spontaneous polarization may lie along any of the cube edges, giving a total 6 possible directions for the spontaneous polarization. Along the direction of spontaneous polarization of a given domain, the material expands, whereas perpendicular to the polarization direction it contracts. Thus, the material is no longer cubic, but corresponds to a so-called tetragonal structure. $BaTiO_3$ has two more transition temperatures: one at 5°C, where the spontaneous polarization changes its direction from one of the cube edges to a direction corresponding to a face diagonal in Fig. 2.22; and one at −80°C where the spontaneous polarization changes from a direction corresponding to a face diagonal to one along a body diagonal. Associated with each of these ferroelectric transitions is a change in the crystal structure of the material. These three transition temperatures are reflected in the dielectric constant and in the spontaneous polarization of the material, as may be seen from Figs. 2.24

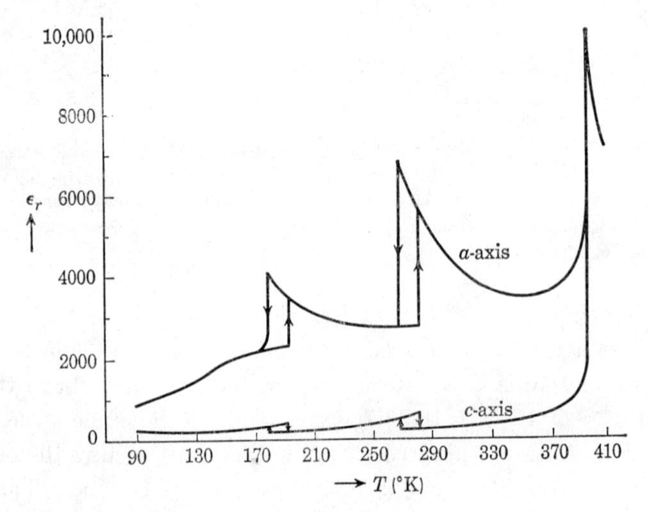


**Fig. 2.24.** The dielectric constant of $BaTiO_3$ as a function of temperature. [After W. J. Merz, *Phys. Rev.* **76,** 1221 (1949)]

and 2.25. The spontaneous polarization represented in Fig. 2.25 was measured along a cube edge over the whole temperature range. Thus, the magnitude in the region between 193°K and 278°K is obtained by multiplying the value given in Fig. 2.25 by $\sqrt{2}$ ($P_s$ in that region is directed along a face diagonal!). Similarly, to obtain the magnitude of $P_s$ in the region below 193°K, one should multiply the value in Fig. 2.24 by $\sqrt{3}$

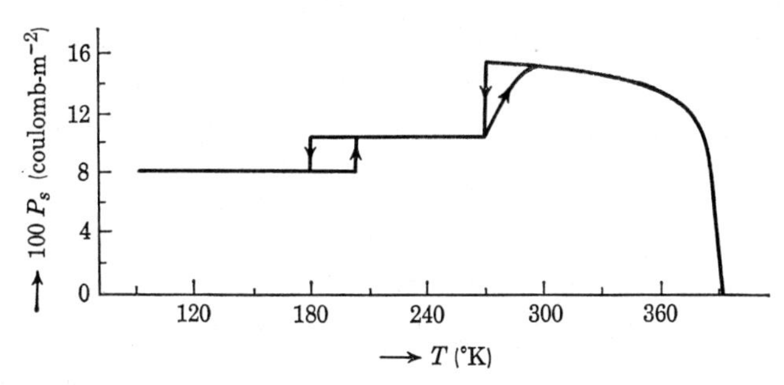


**Fig. 2.25.** The spontaneous polarization of $BaTiO_3$ measured along a cube edge. [After W. J. Merz, *Phys. Rev.* **76,** 1221 (1949)]

($P_s$ directed along body diagonal in this case!). Essentially, therefore, the magnitude of the spontaneous polarization below about 300°K is constant.

## 2.9 Spontaneous polarization

Although we do not intend to give here a detailed discussion of the theory of ferroelectricity, a few remarks should be made with reference to the occurrence of spontaneous polarization. In section 2.7 we noted that in normal solid dielectrics the internal field may be written in the form

$$E_i = E + (\gamma/\epsilon_0)P \tag{2.44}$$

where $E$ is the applied field, $P$ the polarization and $\gamma$ the internal field constant. Although the internal field may be different for different atomic positions in the solid, for the sake of argument we shall assume (2.44) to be valid for all atoms in a hypothetical solid. Suppose this hypothetical solid can be built up by a three-dimensional stacking of units such as the cube in Fig. 2.23 in the case of $BaTiO_3$. Let there be $N$ of these units per m$^3$ and let the total polarizability per unit be $\alpha$. The polarization of the material may then be written as

$$P = N\alpha E_i = N\alpha[E + (\gamma/\epsilon_0)P] \tag{2.45}$$

from which we find upon solving for $P$

$$P = \frac{N\alpha E}{1 - (N\alpha\gamma/\epsilon_0)} \tag{2.46}$$

Does this formula indicate the possibility of spontaneous polarization? What we are asking for is actually this: in the absence of an external field ($E = 0$), does (2.46) allow a non-vanishing value for $P$? The answer is, that such a solution indeed exists, viz. when the denominator in (2.46) equals zero. In other words, if $N\alpha\gamma/\epsilon_0 = 1$, spontaneous polarization is possible. Physically, this means that if the interaction between the atoms is large enough (large $\gamma$), and if $N\alpha$ is large enough, spontaneous polarization may occur. We may also look at this from a somewhat different point of view. Consider the string of atoms in Fig. 2.18 in the absence of an external field. The string may then occur in two states: (i) the atoms are not polarized; i.e., they carry no dipoles; (ii) the atoms induce in each other dipoles such that the string is spontaneously polarized. Which of these two states is realized depends on the energy of the system in the two states. If the energy of the string in the spontaneously polarized state is lower than that of the unpolarized state, the spontaneously polarized state will be the stable configuration of the system; if the energy of the spontaneously polarized string is higher than that of the unpolarized string, the system will not be spontaneously polarized.

Still another way of looking at the condition for spontaneous polarization emerges by writing (2.46) in the form

$$\frac{P}{E} = \frac{N\alpha}{1 - (N\alpha\gamma/\epsilon_0)} = \frac{A}{1 - A\beta} \tag{2.47}$$

where $\beta = \gamma/\epsilon_0$. In this form, the last expression reminds us of the gain of an amplifier with an amplification $A$ and a feedback factor $\beta$ (see Fig. 2.26). If there were no feedback; i.e., if the internal field were the same as

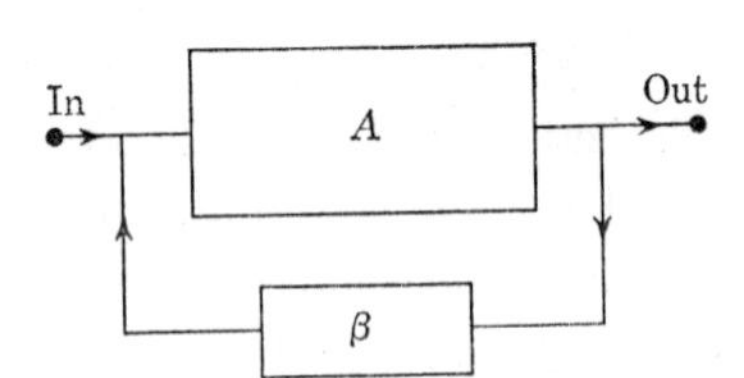


**Fig. 2.26.** Schematic representation of an amplifier with gain $A$ and a feedback loop $\beta$. The total gain is given by expression (2.47).

the applied field, corresponding to $\beta = \gamma = 0$, the "gain" (which in this case corresponds to the polarization per unit field) would be simply $N\alpha = A$. As a result of the feedback associated with the internal field constant $\gamma$, the gain is in general larger than $N\alpha$, and may become infinite if $A\beta = N\alpha\gamma/\epsilon_0 = 1$.

It will be evident from the discussion that a high polarizability $\alpha$ is one of the factors which is favorable for the occurrence of spontaneous polarization. In this connection we remind the reader of a remark made earlier in connection with the high ionic polarizability associated with the $Ti^{4+}$ ion in $BaTiO_3$; this is presumably one of the factors which give rise to the spontaneous polarization in this material.

The next question which arises is this: why does spontaneous polarization usually occur only below a certain temperature? We shall consider here one possible cause for the existence of a Curie temperature. Let us start from equation (2.46), assuming that this expression applies in the nonferroelectric region ($T \geqslant \theta_f$). From the general expression $P = \epsilon_0(\epsilon_r - 1)E$ it then follows that the dielectric constant of the material satisfies the relation:

$$\epsilon_r - 1 = \frac{N\alpha/\epsilon_0}{1 - N\alpha\gamma/\epsilon_0} \tag{2.48}$$

Let us inquire about the temperature dependence of $\epsilon_r$, assuming for simplicity that $\alpha$ and $\gamma$ are temperature independent and that only $N$ is a function of temperature. If $\lambda$ represents the *coefficient of volume expansion* of the material, then we have evidently

$$\frac{1}{N}\frac{dN}{dT} = -\lambda \tag{2.49}$$

The minus sign indicates that as $T$ increases the volume increases and hence $N$ decreases. Now, in "normal" dielectrics for which $\epsilon_r$ is of the order of unity or ten, the influence of expansion of the material with increasing temperature has very little influence on $\epsilon_r$ if $\alpha$ and $\gamma$ are independent of $T$. However, we see from (2.48) that if at a certain temperature $T_1$ the quantity $N(T_1)\alpha\gamma/\epsilon_0$ is only slightly smaller than 1, cooling of the sample may increase $N$ sufficiently to make $N\alpha\gamma/\epsilon_0$ equal to 1 for some temperature $\theta_f$ below $T_1$, with the result that spontaneous polarization may set in at $T = \theta_f$. At the same time, $\epsilon_r - 1$ would be very large in the vicinity of the critical temperature (see Fig. 2.27). These qualitative arguments

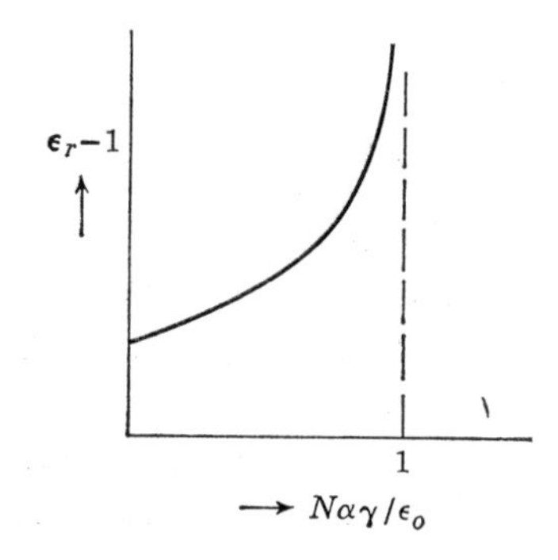


**Fig. 2.27.** Schematic representation of $\epsilon_r - 1$ as a function of $N\alpha\gamma/\epsilon_0$, according to formula (2.48). The value $N\alpha\gamma/\epsilon_0 = 1$ corresponds to the ferroelectric Curie temperature; $N\alpha\gamma/\epsilon_0 < 1$ corresponds to temperatures above the ferroelectric Curie temperature.

show that a Curie temperature may arise in a material of high $\epsilon_r$ simply as a result of the contraction of the material upon cooling.

Assuming the model is correct, how does $\epsilon_r$ vary with temperature in the vicinity of the Curie temperature? Solving for $N$ from (2.48) we find

$$N = \frac{\epsilon_0}{\alpha}\frac{\epsilon_r - 1}{(\epsilon_r - 1)\gamma + 1} \tag{2.50}$$

Differentiating both sides with respect to $T$ and dividing through by $N$ we obtain

$$\frac{1}{N}\frac{dN}{dT} = -\lambda = \frac{d\epsilon_r}{dT}\frac{1}{(\epsilon_r - 1)[(\epsilon_r - 1)\gamma + 1]} \tag{2.51}$$

The temperature coefficient of $\epsilon_r$ is thus related to the coefficient of expansion in accordance with (2.51) if $\gamma$ and $\alpha$ are assumed to be temperature-independent. Now in the vicinity of the critical temperature $\theta_f$, the dielectric constant $\epsilon_r \gg 1$ and since, according to section 2.6, the internal field constant is of the order of unity (or perhaps larger in ferroelectric materials), one may write (2.51) in the approximate form

$$\frac{d\epsilon_r}{dT} \cong -\lambda\gamma\epsilon_r^2 \tag{2.52}$$

How does the dielectric constant near $T = \theta_f$, but in the region $T \geqslant \theta_f$, vary with temperature? Evidently for $T = \theta_f$ we have $N(\theta_f)\alpha\gamma/\epsilon_0 = 1$ so that according to (2.48) $\epsilon_r(\theta_f) = \infty$. Hence we obtain from (2.52)

$$\int_{\epsilon_r=\infty}^{\epsilon_r} \frac{d\epsilon_r}{\epsilon_r^2} = -\lambda\gamma \int_{\theta_f}^{T} dT$$

so that upon integration we obtain

$$\epsilon_r = \frac{1/\lambda\gamma}{T - \theta_f} \quad \text{for} \quad T > \theta_f \tag{2.53}$$

We see that this expression is exactly of the form (2.43); i.e., this model provides us with the experimentally observed Curie-Weiss law for the dielectric constant above the critical temperature (see also Fig. 2.19). Note, however, that in this model the ferroelectric Curie temperature $\theta_f$ is identical with the temperature $\theta$ in the Curie-Weiss law, whereas experimentally $\theta$ is usually a few degrees lower than $\theta_f$.

It is of interest to note that for $BaTiO_3$ the coefficient of expansion $\lambda \cong 3 \times 10^{-5}$ per degree and the observed Curie constant $1/\lambda\gamma \cong 10^5$. Consequently, there is also reasonable agreement as far as the order of magnitude of the Curie constant for $BaTiO_3$ is concerned between the model and experiment, if one assumes $\gamma \approx 1$.

## 2.10 Piezoelectricity

In our previous discussions we have seen that an applied electric field induces dipole moments in atoms or ions, and generally displaces ions relative to each other. Consequently, the dimensions of a specimen undergo slight changes. Mechanical stresses also change the dimensions of a specimen but in general such changes do not produce a dipole moment. In other words, in most materials dielectric polarization produces a mechanical distortion, but a mechanical distortion does not produce polarization. This electromechanical effect, which is present in all materials, is called *electrostriction.* In purely electrostrictive materials, the mechanical deformation produced by a polarization in a given direction is the same as that produced by a polarization in the opposite direction; i.e., mechanical changes can be expressed as a series expansion containing only terms with *even* powers of $P$. A simple example of a material with only electrostrictive properties is given in Fig. 2.28. Note that the basic unit from

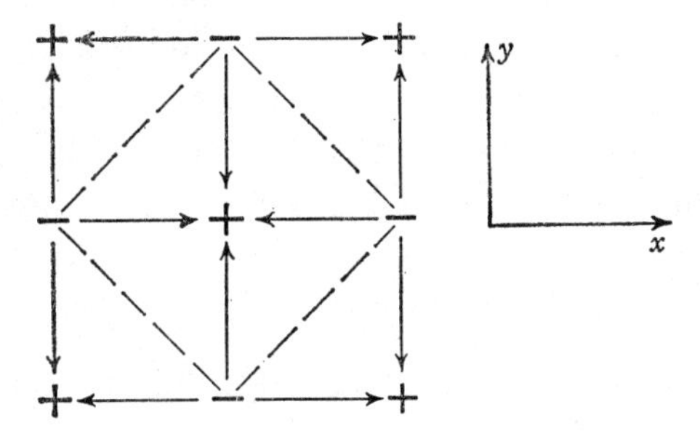


**Fig. 2.28.** A two-dimensional square lattice with only electrostrictive properties. Application of a field along the positive $x$-direction produces the same mechanical deformation as a field along the negative $x$-direction. The dashed square represents the basic unit, and has a center of symmetry.

which this material can be built contains a *center of symmetry;* i.e., starting from the center and drawing a vector to one of the surrounding ions, one finds a similar ion at a position corresponding to a vector of equal length drawn in the opposite direction.

There are solid dielectric materials, however, for which the sign of a mechanical deformation produced by a polarization **P** changes when the direction of the polarization is reversed. Such mechanical deformations then contain at least one term with an *odd* power of $P$. These materials do become polarized upon application of a mechanical stress and are called *piezoelectric;* they are of practical importance because they permit *conversion of mechanical into electrical energy* and vice versa. A two-dimensional example of such a material is represented in Fig. 2.29. The basic unit from which this material can be built evidently *lacks a center of symmetry,* and this is a requirement for a piezoelectric material.

We note that if in Fig. 2.29 we apply a tension along the $x$-direction, the angle $\theta$ will increase, thus giving rise to a polarization in the positive

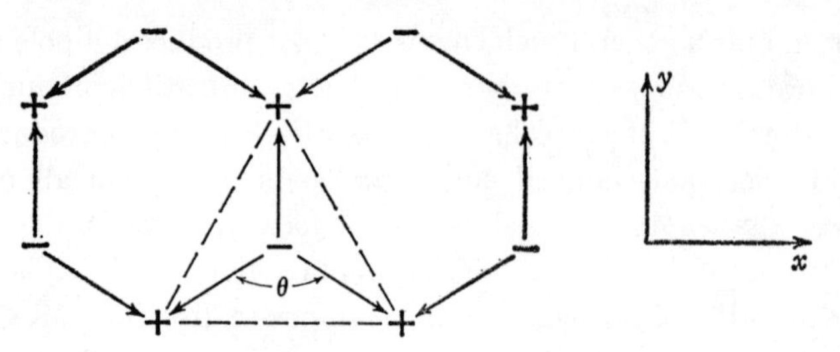


**Fig. 2.29.** A two-dimensional structure with piezoelectric properties; the dashed triangle lacks a center of symmetry.

$y$-direction. But if the unit is compressed along the $x$-direction, $\theta$ will decrease and the polarization will lie along the negative $y$-direction. On the other hand, application of tension or compression to the basic unit in Fig. 2.28 produces no polarization at all, because of the symmetry of the unit. For a detailed discussion of the piezoelectric effect the reader is referred to the books listed below.

## References

C. J. F. Böttcher, *Theory of Electric Polarization,* Elsevier, Amsterdam, 1952.

P. Debye, *Polar Molecules,* Dover, New York, 1945.

A. von Hippel, *Dielectric Materials and Applications,* Technology Press, Cambridge, Massachusetts, and Wiley, New York, 1954.

A. von Hippel, *Dielectrics and Waves,* Technology Press, Cambridge, Massachusetts, and Wiley, New York, 1954.

## Problems

**2.1** A charge of $Q$ coulombs is distributed homogeneously over the surface of a sphere with a radius of $R$ meters; the sphere is in vacuum. Find the flux density $D$, the field strength $E$ and the potential $V$ as a function of the distance $r$ from the center of the sphere for $0 \leqslant r \leqslant \infty$; assume $V(\infty) = 0$.

**2.2** A charge of $Q$ coulombs is distributed homogeneously throughout the volume of a sphere of radius $R$ meters; the sphere is in vacuum. Find the flux density $D$, the field strength $E$ and the potential $V$ as a function of the distance from the center of the sphere for $0 \leqslant R \leqslant \infty$; assume $V(\infty) = 0$.

**2.3** Find the capacitance $C$ of an isolated conducting sphere of radius $R$ meters in vacuum. If the sphere is charged with $Q$ coulombs, what is the energy stored? Assume the potential is zero at infinity.

**2.4** Consider two coaxial metal cylinders of radii $R_1$ and $R_2$. The space between them is filled with a dielectric with a relative dielectric constant $\epsilon_r$. The potential difference applied between the two cylinders is $V$ volts. Find the charge on the cylinders and the capacitance of the system per meter length (neglecting end effects).

**2.5** Electrolytic condensers are manufactured by anodic oxidation of aluminum; the thickness of the aluminum oxide layer formed in this manner is proportional to the anode voltage employed, and amounts to approximately 0.1 micron per 100-volt anode voltage. Find the approximate capacitance for a strip of aluminum of 5 cm by 40 cm, oxidized on both sides to an anode voltage of 500 volts if $\epsilon_r = 8$ for $Al_2O_3$.

**2.6** In the design of oscillator-tank circuits one is frequently faced with the requirement of temperature-independent tuning. If $L$ and $C$ are the equivalent self inductance and capacitance of the circuit, show that this requirement may be expressed mathematically as

$$\frac{1}{L}\frac{dL}{dT} + \frac{1}{C}\frac{dC}{dT} = 0$$

where $T$ is the temperature.

**2.7** A condenser of 1 microfarad contains titanium oxide ($TiO_2$) as a dielectric with $\epsilon_r = 100$. For an applied d-c voltage of 1000 volts, find the energy stored in the condenser as well as the energy stored in polarizing the titanium oxide. Answer the same questions for a 1-microfarad mica condenser, assuming a dielectric constant $\epsilon_r = 5.4$ for mica.

**2.8** Consider a neutral system of point charges, $Q_1, Q_2, \ldots, Q_i, \ldots,$ which are located at the endpoints of a set of vectors $\mathbf{r}_1, \mathbf{r}_2, \ldots, \mathbf{r}_i, \ldots$ drawn from the origin of a coordinate system. Show that the dipole moment $\boldsymbol{\mu} = \sum_i Q_i\mathbf{r}_i$ is independent of the choice of the origin of the coordinate system. (Hint: shift the origin to another point, write down an expression for $\boldsymbol{\mu}$ in the new system, and show that $\boldsymbol{\mu}_{\text{new}} = \boldsymbol{\mu}_{\text{old}}$).

**2.9** With reference to a two-dimensional Cartesian coordinate system $x$, $y$, four point charges are located as follows: a charge of $Q$ coulombs in the point (0, 0); $-Q$ in (1, 0); $2Q$ in (1, 1); and $-2Q$ in (0, 1); the numbers refer to meters. Find the magnitude and direction of the dipole moment of the system.

**2.10** An electrolytic condenser consisting of an oxidized aluminum sheet with an effective surface area of 400 cm$^2$ has a capacitance of 8 microfarads; the dielectric constant of $Al_2O_3$ is $\epsilon_r = 8$. A potential difference of

10 volts is applied between the aluminum and the electrolyte. What is the field strength and what is the total dipole moment induced in the oxide layer?

**2.11** A particle of charge $Q$ coulombs is bound elastically to an equilibrium position with a force constant $f$ newton $m^{-1}$. What is the polarizability of the system?

**2.12** Assuming that the polarizability of an argon atom is equal to $1.43 \times 10^{-40}$ farad $m^2$, calculate the relative dielectric constant of argon at 0°C and 1 atmosphere.

**2.13** According to wave mechanics, an electron in the ground state of a hydrogen atom corresponds to a charge distribution given by expression (1.11). Assuming that the form of the charge distribution remains constant for small applied electric fields, calculate the polarizability of the hydrogen atom by the method used in section 2.3; assume that the displacement of the charge cloud relative to the nucleus $\ll r_1$. Compare the answer with expression (2.12).

**2.14** An atom of polarizability $\alpha$ is placed in a homogeneous field $E$. Show that the energy stored in the polarized atom is equal to $\frac{1}{2}\alpha E^2$.

**2.15** An atom has a polarizability of $10^{-40}$ farad $m^2$; it finds itself at a distance of 10 angstroms from a proton. Calculate the dipole moment induced in the atom and the force with which the proton and the atom attract each other.

**2.16** A sealed-off vessel with two electrodes to measure the dielectric constant of a gas has a pressure of 760 mm of mercury at 300°K. The dielectric constant at 300°K is found to be $\epsilon_r = 1.006715$; at 450°K, $\epsilon_r = 1.005970$. Find the number of molecules in the gas per $m^3$, the dipole moment of the molecules and the polarizability of the molecules.

**2.17** A point dipole of $\mu$ coulomb m finds itself at a distance of $a$ meters from the center of an atom of polarizability $\alpha$ farad $m^2$; the direction of $\boldsymbol{\mu}$ is parallel to the line joining the dipole and the center of the atom. Find the dipole moment induced in the atom.

**2.18** Repeat problem 2.17 for the configuration in which $\mu$ is perpendicular to the line joining the dipole and the center of the atom.

**2.19** The centers of two identical atoms of polarizability $\alpha$ farad $m^2$ are separated by a distance of $a$ meters. A homogeneous electric field $\mathbf{E}$ is applied in a direction parallel to the line joining the centers of the two atoms. Find the internal field $\mathbf{E}_i$ at the position of each of the atoms. If $\alpha = 2 \times 10^{-40}$ farad $m^2$ and $a = 5 \times 10^{-10}$ m, what is the ratio between $E_i$ and $E$?

**2.20** Repeat problem 2.19 for an applied field **E** perpendicular to the line joining the centers of the two atoms.

**2.21** Consider a solid containing $N$ identical atoms per m$^3$; the polarizability of the atoms is $\alpha$ farad m$^2$. Assuming a Lorentz internal field, derive the Clausius-Mosotti relation.

**2.22** A solid contains $5 \times 10^{28}$ identical atoms per m$^3$, each with a polarizability of $2 \times 10^{-40}$ farad m$^2$. Assuming that the internal field is given by the Lorentz formula, calculate the ratio of the internal field to the applied field.

**2.23** Consider the following two infinite arrays of identical equidistance *point* dipoles:

$$\cdots \quad \rightarrow \quad \rightarrow \quad \rightarrow \quad \cdots \quad \rightarrow \quad \cdots \qquad \text{(ferroelectric)}$$
$$\cdots \quad \uparrow \quad \downarrow \quad \uparrow \quad \cdots \quad \downarrow \quad \cdots \qquad \text{(antiferroelectric)}$$

The distance between neighboring dipoles is the same in the two configurations. From an examination of the field produced at the position of a given dipole by all other dipoles in the array, argue which of the two configurations is the more stable one.

# 3

# Behavior of Dielectrics in Alternating Fields

In this chapter we shall discuss the essential aspects of the behavior of dielectric materials when subjected to alternating fields. The discussion is based on the atomic models employed in the preceding chapter, and from the behavior of these models in alternating fields we shall arrive at the frequency dependence of the macroscopic dielectric constant. As a result of the discussion it will become evident that the dielectric constant under these conditions is in general a *complex quantity* of which the imaginary part is a measure for the *dielectric losses* of the material. The discussion in this chapter is by no means complete and serves mainly to illustrate the principles leading to the complex dielectric constant and its interpretation.

## 3.1 Frequency dependence of the electronic polarizability

Let us return at this point to the atomic model employed in section 2.3, and let us inquire what results would be obtained for the polarizability $\alpha_e$ when the model is subjected to an alternating field. For simplicity we shall assume a nucleus of charge $+e$ and a single electron, the latter being represented by an electron cloud of total charge $-e$ distributed homogeneously through the volume of a sphere of radius $R$; the center of the sphere in the absence of an external field coincides with the nucleus (see Fig. 2.5). Since the nucleus is much heavier than the electron cloud, we may consider the nucleus to a good approximation to be at rest, the electron cloud carrying out the motion forced on it by the a-c field. The first problem is, of course, to set up the differential equation which describes the motion of the electron cloud. First consider the following problem: suppose the electron cloud is displaced by an amount $x_0$ relative to the nucleus and then the system is left to itself. What is the differential equation which describes the motion of the electron cloud under these circumstances? From the discussion in section (2.3) it follows that the force which tends to drive the center of the cloud to the nucleus is given by

$$F = -e^2x/4\pi\epsilon_0 R^3 = -ax \tag{3.1}$$

where $x$ is the displacement. The force $F$ is called the *restoring force* and $a$ the *restoring force constant.* Hence, if there were no damping, and in the absence of an applied field, the equation of motion of the electron cloud would be identical with that for a *harmonic oscillator*, viz.,

$$m\frac{d^2x}{dt^2} = -ax \tag{3.2}$$

where $m$ is the mass of the cloud, i.e. the electron mass. It is well known that the solution of (3.2) is

$$x = x_0 \sin(\omega_0 t + \delta) \tag{3.3}$$

where $x_0$ and $\delta$ are integration constants, and where $\omega_0 = (a/m)^{1/2}$ is the *natural* or *resonance angular frequency.* An estimate of the order of magnitude of $\omega_0$ is obtained by putting in (3.1) $R \cong 10^{-10}$ m; this gives with $m = 0.9 \times 10^{-31}$ kg a value of $\omega_0 \cong 10^{16}$ radians sec$^{-1}$. Hence, the frequencies we are talking about lie in the *ultraviolet* part of the electromagnetic spectrum.

Expression (3.2) is incomplete in the sense that it does not take into account the emission of electromagnetic radiation by the system; the emission results from the time variation of the acceleration of the electron cloud and leads to *damping.* In the mechanical case, damping of the oscillating particle would result from the viscosity of the medium in which the particle moves and it is well known that this damping leads to a term proportional to the *velocity* of the particle in the equation of motion. It can be shown that, in the electrical case under consideration, the damping due to radiation may be represented in a similar way; i.e., instead of (3.2) we should write

$$m\frac{d^2x}{dt^2} = -ax - 2b\frac{dx}{dt} \tag{3.4}$$

where the last term is the damping term. The constant $b$ is related to the natural frequency $\omega_0$ in the following manner

$$2b = \mu_0 e^2\omega_0^2/6\pi mc \tag{3.5}$$

where $\mu_0 = 1.257 \times 10^{-6}$ henry m$^{-1}$ is the magnetic permeability of vacuum, and $c = 2.9979 \times 10^8$ m sec$^{-1}$ is the speed of light. Substituting numerical values, the reader may verify that $2b/m \ll \omega_0$, a result which will be used later.

We are now in a position to write down the equation of motion of the electron cloud in the presence of an alternating external field. Let the field be applied in the $x$-direction and let it be represented by $E_0 \cos \omega t$, $\omega$ being the angular frequency. The force on the electron cloud resulting from the field is then $-eE_0 \cos \omega t$ and the equation of motion is

$$m \frac{d^2x}{dt^2} = -ax - 2b\frac{dx}{dt} - eE_0 \cos \omega t \tag{3.6}$$

To solve this equation for $x(t)$ it is convenient to use complex notation. Thus, we shall write*

$$E_0 \cos \omega t = \text{Re}\,[E_0 e^{j\omega t}] = E_0\,\text{Re}\,[e^{j\omega t}] \tag{3.7}$$

and we shall assume the solution to be of the form

$$x(t) = \text{Re}\,[A^* e^{j\omega t}] \tag{3.8}$$

where $A^*$ is in general a complex amplitude. Substituting the last two expressions into (3.6) one obtains

$$\text{Re}\left\{\left[-\omega^2 A^* + \frac{a}{m} A^* + j\frac{2b\omega}{m} A^* + \frac{e}{m} E_0\right] e^{j\omega t}\right\} = 0$$

From this it follows that the expression in square brackets is zero; writing $a/m = \omega_0^2$ in accordance with the definition of the natural frequency $\omega_0$, we find

$$A^* = \frac{(e/m)E_0}{\omega^2 - \omega_0^2 - j(2b/m)} \tag{3.9}$$

What is the induced dipole moment as function of time? In general, $\mu_{ind}(t) = -ex(t)$ so that we find from (3.8) and (3.9)

$$\mu_{ind}(t) = \text{Re}\left[\frac{(e^2/m)E_0 e^{j\omega t}}{\omega_0^2 - \omega^2 + j(2b\omega/m)}\right] \tag{3.10}$$

Since the coefficient of $E_0 e^{j\omega t}$ is a complex quantity, we see that the static definition $\mu_{ind} = \alpha_e E$ cannot be applied in this case. We are therefore compelled to introduce a *complex polarizability* $\alpha_e^*$ by means of the following expression

$$\mu_{ind}(t) = \text{Re}\,[\alpha_e^* E_0 e^{j\omega t}] \tag{3.11}$$

where

$$\alpha_e^* = \frac{e^2/m}{\omega_0^2 - \omega^2 + j(2b\omega/m)} \tag{3.12}$$

* Re [ ] means "real part of [ ]". Complex quantities will be provided with an asterisk superscript.

Writing out the real and imaginary parts of (3.12) we find

$$\alpha_e^* = \frac{e^2}{m}\left[\frac{\omega_0^2 - \omega^2}{(\omega_0^2 - \omega^2)^2 + (4b^2\omega^2/m^2)} - j\frac{2b\omega/m}{(\omega_0^2 - \omega^2)^2 + (4b^2\omega^2/m^2)}\right]$$
$$= \alpha_e' - j\alpha_e'' \tag{3.13}$$

where $\alpha_e'$ and $\alpha_e''$ represent, respectively, the real and imaginary parts of the polarizability.

We shall now discuss the frequency dependence of the real and imaginary parts of $\alpha_e^*$, referring to Fig. 3.1. First of all, we note that for $\omega = 0$

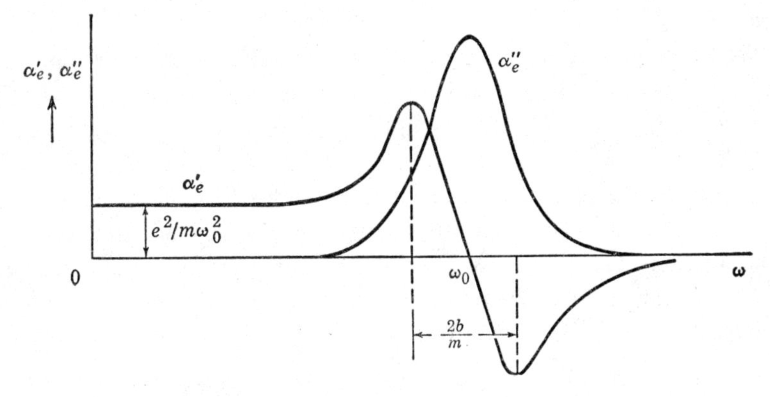


**Fig. 3.1.** Schematic representation of the frequency dependence of the real and imaginary parts $\alpha_e'$ and $\alpha_e''$, respectively, of the electronic polarizability for a single electron.

the imaginary part vanishes, the real part being equal to the *static value* $e^2/m\omega_0^2$. The real part is positive for all values $\omega < \omega_0$ and negative for all values $\omega > \omega_0$; the real part is zero for $\omega = \omega_0$. Remembering that $2b \ll \omega_0$, it is noted that $\alpha_e'$ is essentially constant from zero frequency up to frequencies which become comparable to $\omega_0$. In the region where $\omega$ is nearly equal to $\omega_0$, the behavior can be discussed conveniently by introducing the variable

$$\Delta\omega = \omega_0 - \omega \quad \text{with} \quad \Delta\omega \ll \omega_0$$

We may then write approximately

$$\omega_0^2 - \omega^2 = (\omega_0 + \omega)(\omega_0 - \omega) \cong 2\omega_0\Delta\omega$$

and the real part becomes

$$\alpha_e' \cong \frac{e^2}{m}\frac{2\omega_0\Delta\omega}{4\omega_0^2(\Delta\omega)^2 + 4b^2\omega_0^2/m^2} = \frac{e^2}{m}\frac{(\Delta\omega)/2\omega_0}{(\Delta\omega)^2 + b^2/m^2} \tag{3.14}$$

This expression has a maximum for $\Delta\omega = b/m$ and a minimum for $\Delta\omega = -b/m$, as illustrated in Fig. 3.1. The dampings coefficient $2b$ is thus a measure for the distance between the maximum and minimum in the *dispersion curve* ($\alpha_e'$ versus $\omega$).

The imaginary part $\alpha_e''$ vanishes for $\omega = 0$ as well as for $\omega \to \infty$. In the vicinity of $\omega = \omega_0$, we may write by introducing the variable $\Delta\omega$,

$$\alpha_e'' \cong \frac{e^2b/2m^2\omega_0}{(\Delta\omega)^2 + b^2/m^2} \tag{3.14a}$$

Thus, $\alpha_e''$ exhibits a maximum for $\Delta\omega = 0$, i.e. for $\omega = \omega_0$; the magnitude of the maximum is $(\alpha_e'')_{\max} = e^2/2\omega_0 b$. The width of the bell-shaped curve for $\alpha_e''$ corresponding to half the maximum value is readily found to be $2b/m$.

The consequences of a complex electronic polarizability for the dielectric behavior will be discussed later, but it may be said here already that the *imaginary part of the polarizability gives rise to absorption of energy by the system from the field.*

The model discussed above was limited to the existence of one electron. In general an atom contains a number of electrons, each of them corresponding to a particular force constant $a_i$ and a particular damping constant $b_i$. Consequently, the atom in general will have a series of $\omega_{0i}$ values

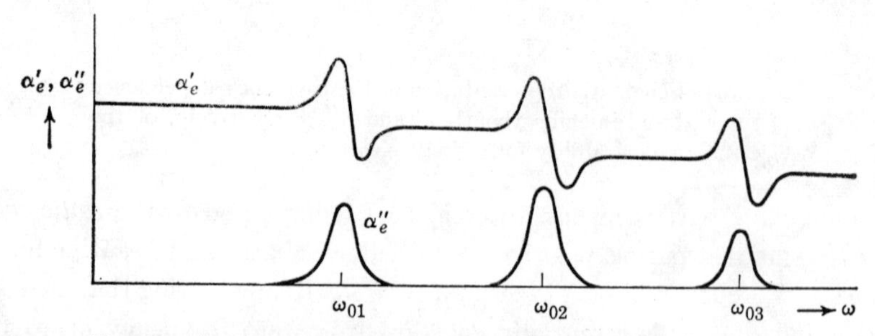


**Fig. 3.2.** Schematic representation of the frequency dependence of the real and imaginary parts of the polarizability of an atom; in this case there are a series of resonance frequencies $\omega_{01}$, $\omega_{02}$, etc.

and the polarizability will exhibit a frequency dependence as indicated schematically in Fig. 3.2.

## 3.2 Ionic polarization as a function of frequency

The frequency dependence of the ionic polarizability can be discussed in complete analogy with the electronic polarizability, the difference between the two cases being of a quantitative nature only. When two ions in a molecule or solid are displaced relative to each other, the restoring force which tends to drive them back to their equilibrium positions is to a good approximation proportional to the displacement. Hence, the forces are harmonic. The masses of the particles in the present case are, of course, those of atoms rather than of electrons. As a result, the natural frequencies of the ionic vibrations lie in the *infrared* part of the electromagnetic spectrum, corresponding to $\omega_0 \approx 10^{14}$ radians per second. Thus, the ionic polarizability of a molecule will also be a complex quantity which may be written in the form

$$\alpha_i^* = \alpha_i' - j\alpha_i'' \tag{3.15}$$

The real part $\alpha_i'$ as a function of the frequency $\omega$ of an applied field exhibits the same features as those represented for $\alpha_e'$ in Fig. 3.2; the only difference is that the $\omega_{0i}$ values for which the maxima and minima occur are now displaced to the infrared region. Similarly, $\alpha_i''$ as a function of frequency will exhibit various bell-shaped maxima, one for each of the characteristic frequencies $\omega_{0i}$.

## 3.3 The complex dielectric constant of non-dipolar solids

On the basis of the information obtained in the preceding two sections, let us consider the frequency dependence of the dielectric constant of a solid, assuming the solid contains no permanent dipoles. The last restriction is not particularly severe in the case of solids, because usually the dipoles are not able to rotate anyway. In the solid state, and also in the liquid state, the applied field must be replaced by the *internal* field $E_i$, as discussed in section 2.6. For simplicity we shall assume that the internal field is given by the Lorentz field (2.33), so that

$$E_i(t) = E(t) + P(t)/3\epsilon_0 \tag{3.16}$$

where $P(t)$ is the electric dipole moment per unit volume at the instant $t$. Let us assume that the solid contains $N$ units per m$^3$ from which the solid may be built up by a three-dimensional stacking. Let each of these units be characterized by an electronic polarizability $\alpha_e^*$ and an ionic polarizability $\alpha_i^*$. In accordance with (3.11) we may then write

$$P(t) = N \operatorname{Re}[(\alpha_e^* + \alpha_i^*)E_{0i}^* e^{j\omega t}] \tag{3.17}$$

where $E_{0i}^*$ is the complex amplitude of the internal field; $\omega$ is the frequency of the applied field. Note that in general, $P(t)$ *will not be in phase with the applied field, or with the internal field.* Consequently, the relationship derived for static fields [see (2.9)]:

$$P = \epsilon_0(\epsilon_r - 1)E$$

is not valid in the case of an alternating field, and we shall therefore define a *complex relative dielectric constant* $\epsilon_r^*$ such that

$$P(t) = \epsilon_0 \operatorname{Re}[(\epsilon_r^* - 1)E_0 e^{j\omega t}] \tag{3.18}$$

where it has been assumed that the applied field is given by $E_0 \cos \omega t$. Note that by introducing the complex $\epsilon_r^*$, we have introduced the possibility of a phase difference between $P(t)$ and $E(t)$. Substituting (3.18) into (3.16) we may then write

$$E_i(t) = \operatorname{Re}\left[\frac{(\epsilon_r^* + 2)}{3} E_0 e^{j\omega t}\right] = \operatorname{Re}[E_{0i}^* e^{j\omega t}] \tag{3.19}$$

Equating (3.17) and (3.18), and substituting (3.19) into (3.17) we find

$$\frac{\epsilon_r^* - 1}{\epsilon_r^* + 2} = \frac{1}{3\epsilon_0} N(\alpha_e^* + \alpha_i^*) \tag{3.20}$$

The reader may compare this result with the Clausius-Mosotti expression (2.38) derived for the static case under the same assumption, viz. a Lorentz field for $E_i$.

The main point of the present discussion is this: the complex dielectric constant of non-dipolar solids is determined by the complex polarizabilities $\alpha_e^*$ and $\alpha_i^*$. Consequently, the behavior of the real and imaginary parts of the polarizabilities as function of frequency will be reflected in the frequency dependence of the dielectric constant. One thus arrives at the conclusion that the real and imaginary parts of $\epsilon_r^*$ defined by

$$\epsilon_r^* = \epsilon_r' - j\epsilon_r'' \tag{3.21}$$

are functions of the frequency of the applied field, and that these functions are determined by $\alpha_e^*(\omega)$ and $\alpha_i^*(\omega)$.

What practical value do these results have for the electrical engineer? This depends on the frequency range in which one happens to be interested. According to the preceding sections, $\alpha_e^*$ and $\alpha_i^*$ are real as long as the frequencies lie below infrared frequencies. Hence, up to microwaves, $\epsilon_r^*$ is essentially real for the materials under discussion and their behavior is the same as it is in static fields. The solids discussed here, however, are idealized in the sense that many of them contain ions which may be displaced over one or more interatomic distances under influence of an external field; this is the case, for example, in glassy materials, and to some extent even in crystalline materials. Such processes may lead to an imaginary part of the dielectric constant, and to dielectric losses as will be seen in subsequent sections.

## 3.4 Dipolar relaxation

So far in this chapter we have discussed only the frequency dependence of the electronic and ionic polarization. From the electrical engineering point of view, the *frequency dependence of the orientational polarization* in liquids and glassy substances is perhaps of greater importance, since it gives rise to dielectric losses in the frequency range between zero and many thousand megacycles, depending upon the substance. Although the discussion refers in particular to permanent dipoles rotating in liquids, the results have much wider applicability.

Consider a liquid containing $N$ permanent dipoles $\mu_p$ per unit volume. Suppose it has been subjected for a long time to a d-c field $E$; let the orientational polarization in equilibrium with the field be $P_o$. When at the instant $t = 0$ the field is suddenly switched off, the polarization will not instantaneously become zero, because there is a certain time required for the rotation of the dipoles. Without going into the details of the molecular processes involved, we shall assume that the polarization as function of time decays to zero in accordance with the formula (see Fig. 3.3a)

$$P_o(t) = P_o e^{-t/\tau} \tag{3.22}$$

The quantity $\tau$ has the dimensions of time and is called the *relaxation time.* In a liquid, $\tau$ increases as the viscosity of the liquid increases, as one would

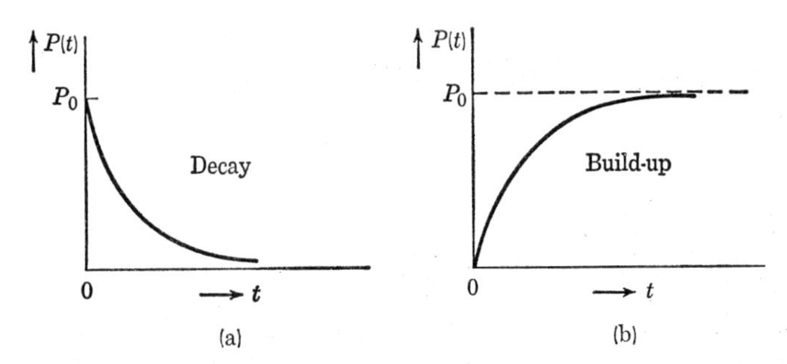


**Fig. 3.3.** Illustrating in (a) the decay of the orientational polarization of a liquid upon switching-off the field at $t = 0$. In (b) a field is switched on at $t = 0$; the curve represents the growth of the orientational polarization with time.

expect from qualitative arguments. The rate of change of the polarization is evidently given by

$$\frac{d}{dt} P_o(t) = -\frac{P_o}{\tau} e^{-t/\tau} = -\frac{P_o(t)}{\tau} \tag{3.23}$$

Note that the smaller the instantaneous value $P_o(t)$, the smaller the rate at which the decay takes place. Since the ultimate value $P_o(\infty) = 0$, we may also write (3.23) as follows:

$$\frac{d}{dt} P_o(t) = \frac{1}{\tau} [P_o(\infty) - P_o(t)] \tag{3.24}$$

The reason for writing the result in this particular form will become obvious later.

Suppose now that an external field has been absent for a long time, and that at $t = 0$ a field $E$ is switched on. What will $P_o(t)$ be during the build-up to the ultimate equilibrium value $P_o$? From analogy with similar physical processes the reader will recognize that the answer to this question is [see Fig. 3.3(b)]

$$P_o(t) = P_o(1 - e^{-t/\tau}) \tag{3.25}$$

The rate of increase is then

$$\frac{d}{dt} P_o(t) = \frac{P_o}{\tau} e^{-t/\tau} = \frac{1}{\tau} [P_o - P_o(t)] \tag{3.26}$$

Since $P_o = P_o(\infty)$ in the present case, note that (3.24) and (3.26) have the same form. In other words, *during build-up as well as during decay*, the rate of change of $P_o(t)$ is, apart from the factor $1/\tau$, equal to the ultimate value corresponding to the field $E$, minus the instantaneous value $P_o(t)$.

The foregoing discussion should be considered as a preparation for the actual question we wish to consider in this section, viz.: given that the equilibrium value of the orientational polarization in a static field $E$ is equal to

$$P_o = \epsilon_0(\epsilon_{r0} - 1)E \tag{3.27}$$

where $\epsilon_{r0}$ is that part of the dielectric constant which measures only the orientational polarization, what is $P_o(t)$ when one applies an alternating field $E_0 \cos \omega t$? In order to answer this question, consider expression (3.24), which was found to hold for decay as well as for build-up, and which we shall now assume to hold also in the case of a-c fields. At the instant $t$, the external field is $E_o \cos \omega t$ and hence at that instant the dipoles are aiming for a $P_o(\infty)$ equal to $\epsilon_0(\epsilon_{r0} - 1)E_0 \cos \omega t$. Consequently, the differential equation for $P_o(t)$ may be written as follows:

$$\frac{d}{dt} P_o(t) = \frac{1}{\tau} [\epsilon_0(\epsilon_{r0} - 1)E_0 \cos \omega t - P_o(t)] \tag{3.28}$$

To solve this equation let us introduce a complex dielectric constant $\epsilon_{r0}^*$ [allowing for the possibility of phase differences between $P_o(t)$ and $E(t)$]

by means of the relation

$$P_o(t) = \epsilon_0 \operatorname{Re}[(\epsilon^*_{r0} - 1)E_0 e^{j\omega t}] \tag{3.29}$$

Substitution into (3.28) then leads to

$$\epsilon^*_{r0} - 1 = \frac{\epsilon_{r0} - 1}{1 + j\omega\tau} \tag{3.30}$$

Hence, for alternating fields, the orientational part of the dielectric constant becomes complex, and is a function of the static value $\epsilon_{r0}$ and of $\omega\tau$. Writing out the real and imaginary parts, we obtain

$$\epsilon^*_{r0} - 1 = (\epsilon_{r0} - 1)\left[\frac{1}{1 + \omega^2\tau^2} - j\frac{\omega\tau}{1 + \omega^2\tau^2}\right] \tag{3.31}$$

For the polarization we find from (3.29) and (3.31)

$$P_o(t) = \frac{\epsilon_0(\epsilon_{r0} - 1)}{1 + \omega^2\tau^2} E_0 \cos\omega t + \frac{\epsilon_0(\epsilon_{r0} - 1)\omega\tau}{1 + \omega^2\tau^2} E_0 \sin\omega t \tag{3.32}$$

Note that the first term on the right-hand side is in phase with the applied field, whereas the second term lags by 90 degrees. The frequency dependence of the in-phase and the out-of-phase components are represented in Fig. 3.4. It is observed that the in-phase component of $P_o(t)$ begins to dis-

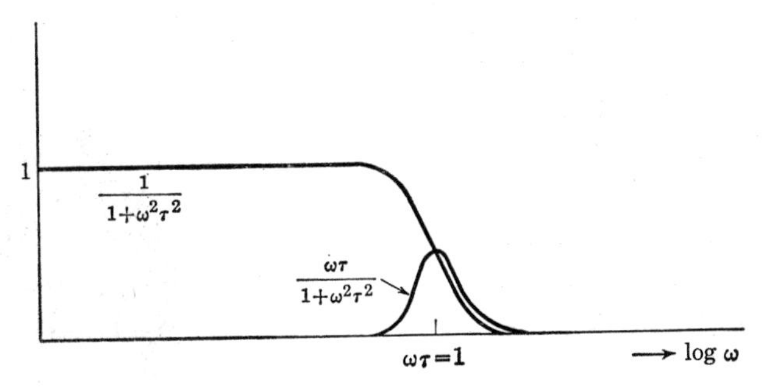


**Fig. 3.4.** A schematic plot of $1/(1 + \omega^2\tau^2)$ and $\omega\tau/(1 + \omega^2\tau^2)$ as a function of the logarithm of the frequency.

appear when $\omega\tau$ becomes comparable to unity. When $\omega\tau \gg 1$, the dipoles cannot follow the field variations and hence the polarization vanishes. The out-of-phase component of $P_o(t)$ has the same bell shape as the imaginary part of the electronic and ionic polarizabilities, and is a measure for the absorption of energy as we shall see in the next section. By way of illustration we give here some values for $\tau$ derived from dielectric measurements for ice and propyl alcohol at various temperatures.

*Propyl Alcohol:*

| *Temp.* (°C): | 20° | 0 | −20° | −40° | −60° |
|---|---|---|---|---|---|
| $10^{10}\tau$ (sec): | 0.9 | 1.6 | 3.2 | 7.4 | 26 |

*Ice:*

| *Temp.* (°C): | −5° | −22° |
|---|---|---|
| $10^{6}\tau$ (sec): | 2.7 | 18 |

Note that as the temperature is reduced, the relaxation time increases and the frequency for which $\omega\tau = 1$ decreases. Note also that the relaxation time is about $10^4$ times as long in ice as in liquid propyl alcohol.

The differential equation for $P_o(t)$, (3.28), also applies to the following situation: With reference to Fig. 3.5 suppose an ion in a particular solid

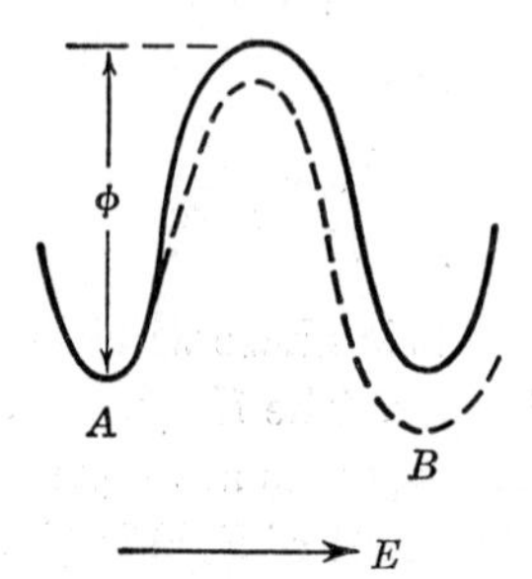


**Fig. 3.5.** The fully drawn curve represents the potential energy of a positive ion as function of a coordinate which coincides with the line joining two possible positions for the ion, $A$ and $B$. The dashed curve corresponds to the potential energy in the presence of a field as indicated.

can occupy two positions $A$ and $B$ of equal energy, the two positions being separated by a potential energy barrier $\phi$. In the absence of an external field the probability that the particle will be found in $A$ is the same as that for $B$. When an external field is applied, position $B$ may be preferred if the energy of the particle is lower there than when it resides in $A$. For a solid containing a significant number of such ions, the process of the ions "jumping" into preferred $B$-sites may contribute appreciably to the dielectric constant. In an alternating field, these ions will contribute in accordance with the formulas derived above, and $\tau$ must then be interpreted as the average time required for an ion to jump from $A$ to $B$. It can be shown that jumping times of this kind depend on the potential barrier $\phi$ and on the temperature as follows:

$$\tau = Ae^{\phi/kT} \tag{3.33}$$

where $A$ is a constant; $k$ is Boltzmann's constant ($= 1.38 \times 10^{-23}$ joule per degree. Thus, as $T$ increases, $\tau$ decreases, as one would expect from qualitative arguments. In non-crystalline materials such as glasses, it is quite likely that there exists a variety of potential barriers $\phi$, and hence a variety of $\tau$-values. Evidence for this will be presented in the next section.

## 3.5 Dielectric losses

In the preceding sections we have discussed the frequency dependence of the electronic, ionic and orientational contributions to the polarization. Since these contributions are additive, a material may be characterized by a complex dielectric constant

$$\epsilon_r^* = \epsilon_r' - j\epsilon_r'' \tag{3.34}$$

in which the real and imaginary parts $\epsilon_r'$ and $\epsilon_r''$ incorporate all three contributions. In the present section we shall show that the imaginary part gives rise to absorption of energy by the material from the alternating field. For this purpose consider a parallel plate condenser filled with a material characterized by $\epsilon_r^*$; the functions $\epsilon_r'(\omega)$ and $\epsilon_r''(\omega)$ are assumed to be given. Let the applied alternating voltage produce a field $E_0 \cos \omega t$. Suppose that at a given instant the charge per unit area on the plates is $\pm Q(t)$. Since the flux density is numerically equal to the charge density, we must have $D(t) = Q(t)$. Also, since the current density is equal to $J(t) = dQ/dt$ we may write

$$J(t) = \frac{dD}{dt} \tag{3.35}$$

On the other hand, since $E(t) = \text{Re}\,[E_0 e^{j\omega t}]$ we may write in accordance with the meaning of the complex dielectric constant

$$D(t) = \text{Re}\,[\epsilon_0 \epsilon_r^* E_0 e^{j\omega t}] = \epsilon_0 E_0 \,\text{Re}\,[\epsilon_r^* e^{j\omega t}] \tag{3.36}$$

Substituting (3.34) into this expression we find for the current density from (3.35)

$$\begin{aligned} J(t) &= \epsilon_0 E_0 \,\text{Re}\,[(\epsilon_r' - j\epsilon_r'')j\omega e^{j\omega t}] \\ &= \omega\epsilon_0 E_0[\epsilon_r'' \cos \omega t - \epsilon_r' \sin \omega t] \end{aligned} \tag{3.37}$$

Note that the imaginary part $\epsilon_r''$ of the dielectric constant determines the component of the current which is in phase with the applied field. Also, the real part of the dielectric constant, $\epsilon_r'$, is coupled with a time factor which is 90 degrees out of phase with the applied field. The reader will readily recognize that, on the average, the last term in (3.37) does not give rise to absorption of energy, whereas the term containing $\epsilon_r''$ does. The instantaneous power per m$^3$ absorbed by the medium is given by $J(t)E(t)$; hence, each second the material absorbs an amount of energy per m$^3$ given by

$$W(t) = \frac{1}{2\pi} \int_0^{2\pi} J(t)E(t)\, d(\omega t) \tag{3.38}$$

Substituting $J(t)$ from (3.37) one readily finds

$$W(t) = (\omega/2)\epsilon_0\epsilon_r''E_0^2 \tag{3.39}$$

Thus, the *absorption of energy is proportional to the imaginary part of the complex dielectric constant;* whenever there is energy dissipated in the medium we speak of *dielectric losses.*

It follows from (3.37) that a condenser containing a lossy dielectric may be represented by an equivalent circuit which consists of a pure capacitance and a parallel resistance, the latter being inversely proportional to $\epsilon_r''\omega$. It is customary to characterize the losses of a dielectric at a certain frequency and temperature by the so-called "*loss-tangent*," tan $\delta$, defined as

$$\tan \delta = \epsilon_r''/\epsilon_r' \tag{3.40}$$

The physical meaning of the angle $\delta$ may be derived from expression (3.37). If there are no losses, $\epsilon_r'' = 0$ and the current density is then given by $\omega\epsilon_0 E_0 \epsilon_r' \cos(\omega t + 90°)$; i.e., the current leads the field by 90 degrees. Under these circumstances $\delta = 0$. If there is a current component in phase with the field, the resulting current will no longer lead the field by 90 degrees but by $90° - \delta$, as indicated in Fig. 3.6.

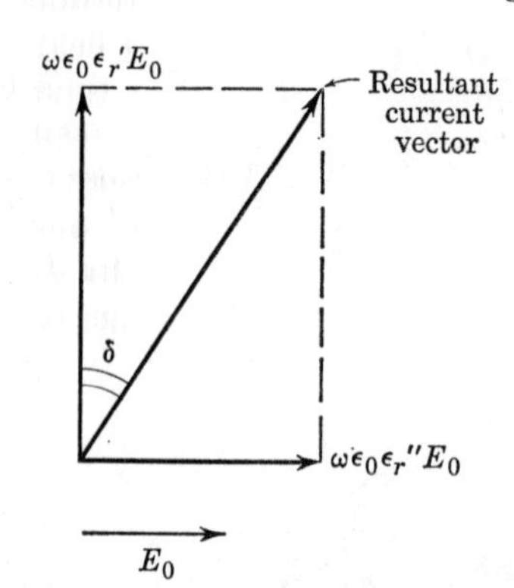


**Fig. 3.6.** Illustrating the vector relationship between the field vector $E_0$, the current vector $\omega\epsilon_0\epsilon_r'E_0$ which leads the field by 90 degrees and the current vector $\omega\epsilon_0\epsilon_r''E_0$ which is in phase with the field. The loss angle $\delta$ is indicated.

The dielectric losses in the radio frequency region are usually due to dipole rotation or to ions jumping from one equilibrium position to another. Losses in this region may also be due to a small degree of d-c *conductivity* of the material, but this subject will not be discussed here. The dielectric losses associated with the ionic vibrations, the frequencies of which fall in the infrared region, are usually referred to as *infrared absorption.* Similarly, the losses in the optical region, associated with the electrons, are referred to as *optical absorption.* The occurrence of absorption in the optical region is the source of the color of materials. For example, a crystal of NaCl is transparent in the visible region; this means that there is negligible absorption for the corresponding frequencies. However, after the NaCl has

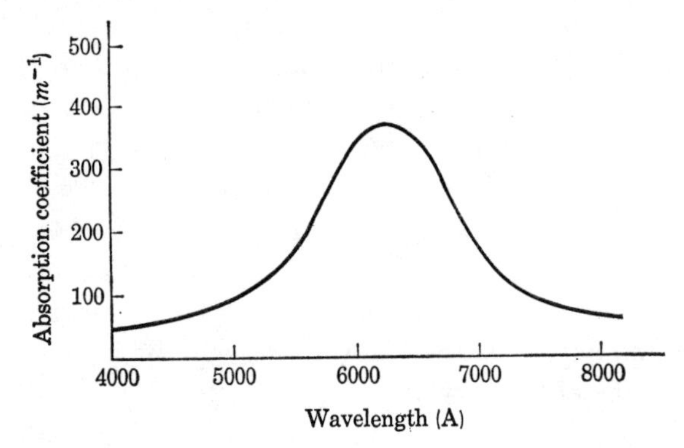


**Fig. 3.7.** The optical absorption as a function of wavelength resulting from *F*-centers in a KBr crystal, at room temperature. [After A. von Hippel, E. P. Gross, T. G. Telatis, and M. Geller, *Phys. Rev.* **91,** 568 (1953)]

been exposed to X-rays, one finds that it has turned yellow-brown. The reason for this is that, after irradiation with X-rays, a relatively small number of electrons, which have been transferred to positions in the lattice

**Table 3.1.** THE REAL PART OF THE RELATIVE DIELECTRIC CONSTANT, $\epsilon_r'$, AND THE LOSS TANGENT OF VARIOUS DIELECTRICS AT A NUMBER OF FREQUENCIES. (Selected from A. von Hippel, *Dielectric Materials and Applications*)

| Material | | Frequency in cycles per second: $10^2$ | $10^4$ | $10^6$ | $10^8$ | $3 \times 10^9$ |
|---|---|---|---|---|---|---|
| Pyranol 1467 | $\epsilon_r'$ | 4.42 | 4.40 | 4.40 | 4.08 | 2.84 |
| | $10^4 \tan \delta$ | 36 | 4 | 25 | 1300 | 1200 |
| Cable oil 5314 | $\epsilon_r'$ | 2.25 | 2.25 | | | 2.22 |
| | $10^4 \tan \delta$ | 3 | 0.4 | | | 18 |
| Teflon | $\epsilon_r'$ | 2.1 | 2.1 | 2.1 | 2.1 | 2.1 |
| | $10^4 \tan \delta$ | 5 | 3 | 2 | 2 | 1.5 |
| Polystyrene | $\epsilon_r'$ | 2.59 | 2.56 | 2.56 | 2.55 | 2.55 |
| | $10^4 \tan \delta$ | 0.5 | 0.5 | 0.7 | 1.0 | 3.3 |
| Polyethylene | $\epsilon_r'$ | 2.25 | 2.25 | 2.25 | 2.25 | 2.25 |
| | $10^4 \tan \delta$ | 5 | 3 | 4 | | 3 |
| Nylon 66 | $\epsilon_r'$ | 3.88 | 3.60 | 3.33 | 3.16 | 3.03 |
| | $10^4 \tan \delta$ | 144 | 233 | 257 | 210 | 128 |
| Bakelite BM-120 | $\epsilon_r'$ | 4.87 | 4.62 | 4.36 | 3.95 | 3.70 |
| | $10^4 \tan \delta$ | 300 | 200 | 280 | 380 | 438 |
| Glass (Corning 0010) | $\epsilon_r'$ | 6.68 | 6.57 | 6.43 | 6.33 | 6.1 |
| | $10^4 \tan \delta$ | 77 | 35 | 16 | 23 | 60 |
| Porcelain No. 4462 | $\epsilon_r'$ | 8.99 | 8.95 | 8.95 | 8.95 | 8.90 |
| | $10^4 \tan \delta$ | 22 | 6.0 | 2.0 | 4.0 | 11 |

where they are not bound so strongly as they were before, give rise to resonance frequencies lying in the visible part of the spectrum. When white light passes through the crystal, a fraction of the light corresponding to a narrow frequency region is absorbed, and the transmitted light is therefore colored. The centers which are responsible for this particular type of absorption are called F-centers (Farbe is the German word for color); they consist of electrons occupying positions in which negative ions are missing. This type of *color center* occurs in all alkali halides as well as in other ionic crystals. An example of F-center absorption in KBr is given in Fig. 3.7; note the bell shape of the curve.

We finally give in Table 3.1 values for the real part of the dielectric constant, $\epsilon_r'$, and for tan $\delta$ for various materials at a number of frequencies. For a collection of data for a large number of materials, the reader is referred to *Dielectric Materials and Applications*, edited by A. R. von Hippel and cited at the end of the preceding chapter.

## References

See those given at the end of Chapter 2.

## Problems

**3.1** (a) Consider a gas containing $N$ similar atoms per m³ of a polarizability $\alpha$. On the basis of expression (3.10) for the induced dipole moment resulting from an alternating field, show that the dielectric constant of the gas is given by

$$\epsilon_r^* = 1 + \frac{Ne^2/m\epsilon_0}{\omega_0^2 - \omega^2 + j2b\omega/m}$$

(b) Consider two parallel metal plates with a separation of 1 m. The space between the plates is occupied by the gas referred to under (a). Show that the admittance per m² plate area of this condenser is given by

$$Y^* = j\omega\left[\epsilon_0 + \frac{Ne^2/m}{\omega_0^2 - \omega^2 + j2b\omega/m}\right]$$

(c) Consider the circuit in the figure. Show that the admittance of this

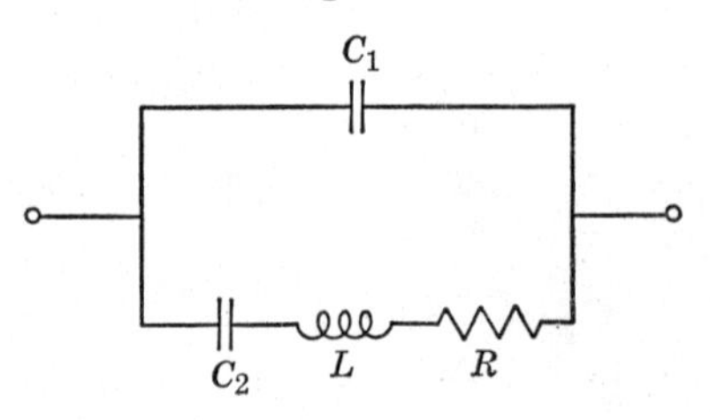

circuit is given by

$$Y^* = jC_1\omega + \frac{j\omega/L}{\omega_0^2 - \omega^2 + jR\omega/L}$$

where $\omega_0^2 = 1/LC_2$.

(d) Compare the results obtained under (b) and (c), and show that the condenser mentioned under (b) has an equivalent circuit as indicated under (c) with the following identifications:

$$C_1 = \epsilon_0; \quad L = m/Ne^2; \quad R = 2b/Ne^2; \quad C_2 = Ne^2/m\omega_0^2$$

**3.2** (a) Consider a parallel arrangement of a capacitance $C$ and a resistance $R$. An external voltage $V(t) = V_0 \cos \omega t$ is applied to this arrangement. Show that the total current $i(t)$ is given by

$$i(t) = (V_0/R) \cos \omega t - C\omega V_0 \sin \omega t$$

(b) Consider a parallel plate condenser with a lossy dielectric between them. At an angular frequency $\omega$ let the dielectric be characterized by a complex dielectric constant $\epsilon_r^* = \epsilon_r' - j\epsilon_r''$. The area of the plates is 1 m$^2$, the distance between them 1 m. For an applied voltage $V(t) = V_0 \cos \omega t$ show that the current through the lossy condenser is given by

$$i(t) = (\epsilon_0\epsilon_r''\omega V_0) \cos \omega t - (\epsilon_0\epsilon_r' V_0\omega) \sin \omega t$$

(c) Compare the results obtained under (a) and (b) and note the occurrence of current components in phase and out of phase with the applied field. Show that the lossy condenser can be represented by an equivalent circuit consisting of a parallel $R$-$C$ arrangement with

$$R = 1/\epsilon_0\epsilon_r''\omega \qquad \text{and} \qquad C = \epsilon_0\epsilon_r'$$

(d) What is the loss tangent of the condenser in (b) expressed in terms of the equivalent $R$ and $C$?

(e) Are the elements of the equivalent circuit independent of the frequency?

**3.3** (a) Suppose a dielectric has a complex dielectric constant given by $\epsilon_r^* = \epsilon_{ei} + \epsilon_{r0}^*$ where $\epsilon_{r0}^*$ refers to the dipole orientations and $\epsilon_{ei}$ is a real quantity referring to the electronic and ionic polarizations. Assume that $\epsilon_{r0}^*$ is determined by a simple relaxation time $\tau$, as in formula (3.30). Consider the space between two parallel metal plates filled with this dielectric. If the distance between the plates is 1 m, show that the admittance of the condenser per m$^2$ plate area is equal to

$$Y^* = j\omega\epsilon_0 \left[\epsilon_{ei} + 1 + \frac{\epsilon_{r0} - 1}{1 + j\omega\tau}\right]$$

(b) Consider the circuit in the figure. Show that the admittance of this

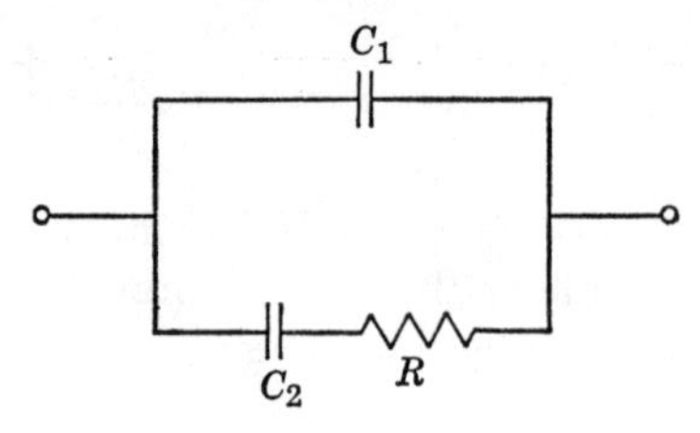


circuit is equal to

$$Y^* = j\omega\left[C_1 + \frac{C_2}{1 + j\omega\tau}\right]$$

where $\tau = RC_2$.

(c) From a comparison of the results obtained under (a) and (b) show that the circuit is the equivalent of the condenser under (a) with the following identification:

$$C_1 = \epsilon_0(\epsilon_{ei} + 1); \quad C_2 = \epsilon_0(\epsilon_{r0} - 1); \quad R = \tau/[\epsilon_0(\epsilon_{r0} - 1)]$$

**3.4** A parallel plate condenser has an area of 10 $cm^2$ and a separation of 0.1 mm. The space between the plates is filled with polyethylene. An alternating voltage with an amplitude of 2 volts is applied at a frequency of 1 megacycle. Given that at this frequency the real part of the relative dielectric constant is 2.25 and the loss tangent is $4 \times 10^{-4}$, find the elements of an equivalent parallel *R-C* circuit. Also calculate the energy dissipation per second.

**3.5** For a polar liquid, make a qualitative sketch of the real and imaginary parts of the dielectric constant at two temperatures as a function of the frequency of an applied radio frequency field.

**3.6** For a polar liquid, make a qualitative sketch of the real and imaginary parts of the dielectric constant as a function of temperature at a given radio frequency.

# 4

# Magnetic Properties of Materials

In this chapter we attempt to explain the differences between the various types of magnetic materials in terms of the magnetic properties of atoms and the interactions among these atoms. The chapter is divided into two parts. Part I is intended to refresh the reader's memory on some fundamental concepts concerning magnetic fields, and to illustrate the essence of the atomic theory of magnetic dipoles with reference to simple models. In part II the information gathered in the first part is used to discuss the atomic interpretation of dia-, para-, ferro-, antiferro- and ferrimagnetism.

## *Part I. Preparatory Discussion*

### 4.1 Summary of concepts pertaining to magnetic fields

In this section the reader is reminded of some fundamental concepts which are discussed in detail in courses on magnetic fields. The *magnetic flux density* in a point of space is denoted by a vector **B**. In the mks system, the unit of flux density may be defined in terms of the force exerted by a magnetic field on a current-carrying wire. Consider, in Fig. 4.1, an element $dl$ of a wire carrying a current of $I$ amperes; in a magnetic field of flux density **B**, the force on the element $dl$ is given by

$$d\mathbf{F} = \mathbf{I} \times \mathbf{B}\, dl \tag{4.1}$$

Thus, the direction of the force $d\mathbf{F}$ is perpendicular to the vectors **I** and **B**, and coincides with the direction in which a right-handed screw advances

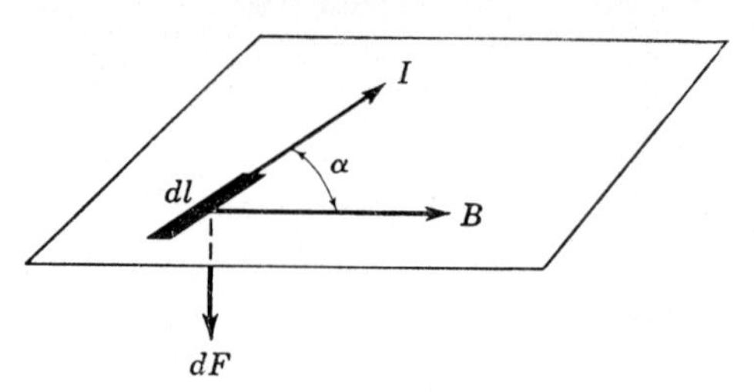


**Fig. 4.1.** Illustrating the force $dF$ exerted by a magnetic flux density $\mathbf{B}$ on an element of wire $dl$ carrying a current $\mathbf{I}$.

when rotated from **I** to **B**. The magnitude of the force is equal to

$$dF = IB\,dl \sin \alpha \tag{4.2}$$

where $\alpha$ is the angle between **I** and **B** as indicated in Fig. 4.1. Since the proportionality constant in (4.1) has been chosen equal to unity, the units of $B$ are fixed by the units of $F$ (in newtons), $I$ (in amperes), and $dl$ (in meters). Thus, $B$ is expressed in newton amp$^{-1}$ m$^{-1}$. One usually calls

$$1 \text{ newton amp}^{-1}\text{ m}^{-1} \equiv 1 \text{ weber m}^{-2} \tag{4.3}$$

Magnetic fields are produced by electric currents; the magnetic flux density produced in a given point by such currents is governed by the law of Biot and Savart. With reference to Fig. 4.2, consider an element $dl$ of

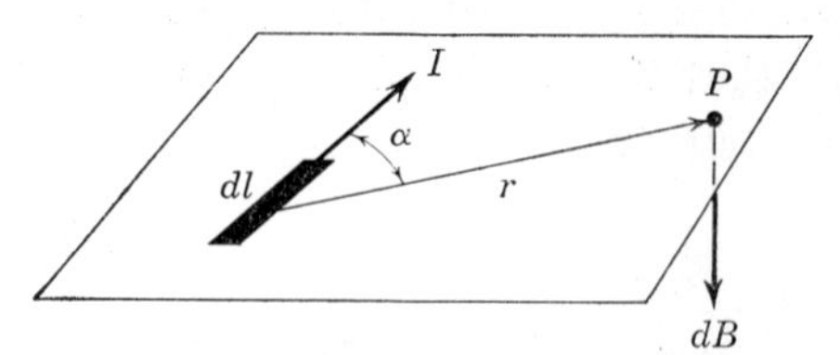


**Fig. 4.2.** Illustrating the contribution to the flux density, $d\mathbf{B}$, resulting from a current-carrying element $dl$.

a wire carrying a current $I$ as indicated. The flux density contributed by this element in a point $P$, located at the end of a vector **r** as indicated, is given by

$$d\mathbf{B} = \frac{\mu_0 \mu_r\, dl}{4\pi r^3} \mathbf{I} \times \mathbf{r} \tag{4.4}$$

Here, $\mu_0$ is usually referred to as the *permeability of free space;* it is numerically equal to $4\pi \times 10^{-7} = 1.257 \times 10^{-6}$ henry m$^{-1}$ (or weber m$^{-1}$ amp$^{-1}$). The quantity $\mu_r$ is the *relative permeability* of the medium; it is a pure number which is equal to unity for vacuum. At this point we may remark that $\mu_0$ (as $\epsilon_0$ in the dielectric case) has no physical significance other than that it appears as a result of the particular system of units used here. The quantity $\mu_r$ (as $\epsilon_r$ in the dielectric case) is the only parameter which can be interpreted in terms of the *atomic* properties of the medium. According to (4.4), the direction of the flux density $d\mathbf{B}$ is perpendicular to the vectors $\mathbf{I}$ and $\mathbf{r}$, and coincides with the direction in which a right-handed screw advances when rotated from $\mathbf{I}$ to $\mathbf{r}$. The magnitude of the flux density contributed by the element $dl$ in Fig. 4.2 is

$$dB = \frac{\mu_0\mu_r}{4\pi r^2} I\, dl \sin\alpha \tag{4.5}$$

where $\alpha$ is the angle between $\mathbf{I}$ and $\mathbf{r}$ as indicated.

As a particular application of the law of Biot and Savart, we leave it up to the reader to show that the magnitude of the flux density produced in a point $P$ by an infinitely long wire carrying a current $I$ is given by

$$B_P = \mu_0\mu_r I/2\pi a \tag{4.6}$$

where $a$ is the distance to point $P$ from the axis of the wire.

In the mks system, the units of the *magnetic field intensity*, $\mathbf{H}$, are determined from the notion that the line integral of $\mathbf{H}$ along a closed curve is equal to the total current enclosed. Thus, with reference to Fig. 4.3 we write

$$\oint \mathbf{H} \cdot d\mathbf{l} = I \tag{4.7}$$

where $I$ represents the current in amperes enclosed by the curve chosen. Thus, the magnetic field intensity $H$ is expressed in amperes $\mathrm{m}^{-1}$. Apply-

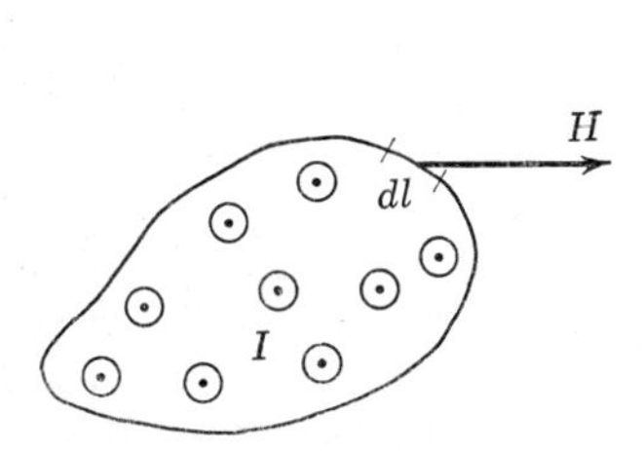


**Fig. 4.3.** The line integral of **H** along the closed curve is equal to the total current $I$ enclosed by the curve; the current in this case flows into the paper.

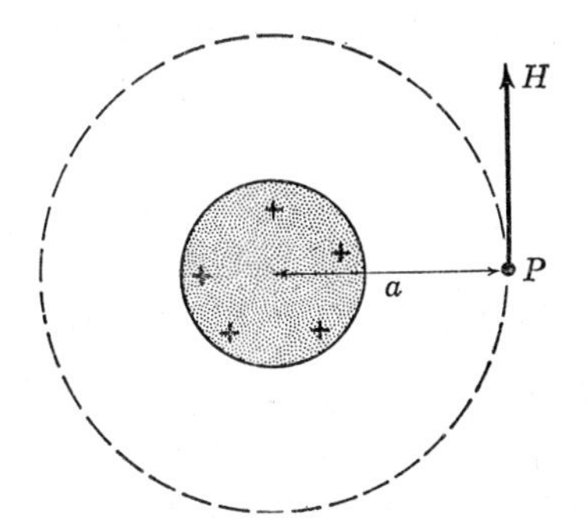


**Fig. 4.4.** Cross section through a wire carrying a current flowing out of the paper; the magnetic field produced in point $P$ is indicated.

ing this to the case of an infinite wire carrying a current $I$, let us calculate the magnetic field intensity in a point $P$ at a distance $a$ from the axis of the wire. Because of the cylindrical symmetry of the problem at hand, we choose as the closed path a circle of radius $a$ (see Fig. 4.4). Then, since **H** is tangent everywhere along the circle, we simply have

$$\oint \mathbf{H} \cdot d\mathbf{l} = 2\pi a H = I$$

or

$$H = I/2\pi a \tag{4.8}$$

Since for the same problem, the flux density is given by (4.6), we arrive at the well-known relation between B and H:

$$\mathbf{B} = \mu_0 \mu_r \mathbf{H} \tag{4.9}$$

In this derivation, we have assumed tacitly that **B** and **H** are parallel vectors; i.e., we have assumed an isotropic medium. We have also assumed that a relative permeability $\mu_r$ can be defined for the material in question. This implies that there exists a unique relationship between **B** and **H** in the material, which excludes ferromagnetic materials; the properties of the latter will be discussed in later sections.

## 4.2 The magnetic dipole moment of a current loop

An essential difference between magnetism and electricity is that in the latter we encounter separate positive and negative charges, whereas in magnetism there are no separate positive and negative poles. This is a consequence of the interpretation of magnetic fields in terms of the motion of electric charges. In the present section we remind the reader of the fact that a current loop produces, at large distances, a magnetic field which is identical with that of a *magnetic dipole moment;* proofs of this statement can be found in textbooks on field theory. In order to illustrate the equivalence of a current loop and a magnetic dipole, we choose an example which is particularly suitable for the subject matter to be discussed in subsequent sections; although we shall consider a simple case, the result is of general validity.

In Fig. 4.5 consider a rectangular wire carrying a current $I$ as indicated; the plane of the rectangle is perpendicular to the paper. We further assume the presence of a homogeneous magnetic flux density **B**, and consider the forces acting on the current-carrying parts of the rectangle. Making use of expression (4.1) one finds for the magnitude of the forces exerted on the elements $PS$ and $QR$

$$F = (PS)IB \tag{4.10}$$

The directions of the two forces are as indicated in Fig. 4.5. It also follows

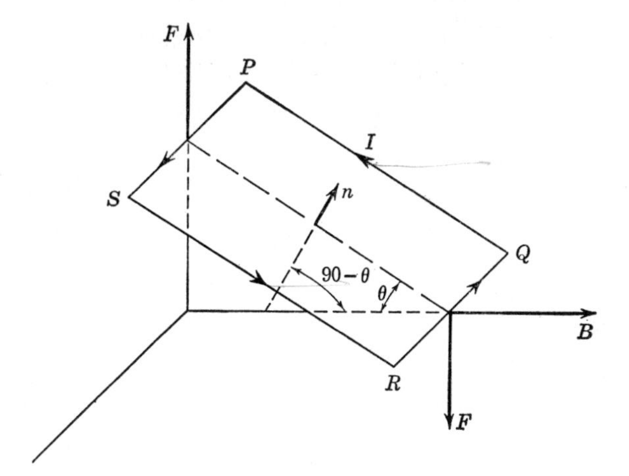


**Fig. 4.5.** Rectangular current loop $PQRS$, carrying a current $I$ and subjected to a homogeneous flux density **B**. The resultant forces on the rectangle are indicated by $F$.

from (4.1) that the forces exerted on the elements $PQ$ and $RS$ will cancel each other. Hence, the flux density exerts a torque $T$ on the rectangle, tending to rotate it to the right, equal to

$$IB(PS)(PQ)\cos\theta \tag{4.11}$$

where $\theta$ is the angle indicated. Denoting the area of the loop by $A$, we may write

$$T = IBA\cos\theta \tag{4.12}$$

For later comparison, let **n** denote a unit vector in the direction normal to the rectangle, and pointing upwards, as indicated in Fig. 4.5. The direction of **n** is the same as that in which a right-hand screw would advance when rotated in the direction of the current flow. Since the angle between **n** and **B** is $(90° - \theta)$, we may write

$$\mathbf{T} = IA\mathbf{n} \times \mathbf{B} \tag{4.13}$$

where the magnitude of the cross-product is equal to $B \sin(\mathbf{n}, \mathbf{B}) = B\cos\theta$. At this point, we remind the reader that the torque produced by an electric field **E** on an electric dipole $\boldsymbol{\mu}$ is equal to $\boldsymbol{\mu} \times \mathbf{E}$ (see section 2.5). We thus see that, apart from a constant, expression (4.13) is indistinguishable from that for the torque exerted on a *magnetic dipole moment with its direction along the unit vector* **n**. Although other choices are possible with regard to the units in which one wishes to express magnetic dipole moments, we shall define the magnetic dipole moment $\boldsymbol{\mu}_m$ associated with the current loop as

$$\boldsymbol{\mu}_m = \mathbf{n}IA \tag{4.14}$$

Thus, $\mu_m$ is expressed in ampere m², and the torque on a magnetic dipole produced by a flux density **B** is according to (4.14) and (4.13) given by

$$\mathbf{T} = \boldsymbol{\mu}_m \times \mathbf{B} \tag{4.15}$$

Although we have derived the equivalence of a current loop and a magnetic dipole moment for a special geometrical form of the loop, it can be shown that the results apply for a current loop of any shape.

## 4.3 The magnetization from a macroscopic viewpoint

In the macroscopic description of electric fields, we encountered three vectors: the flux density **D**, the field intensity **E**, and the polarization **P**; the latter represents the electric dipole moment per unit volume in the material. In section 2.2 we derived a relationship between **P** and **E**, leading to the formula

$$\mathbf{P} = \epsilon_0(\epsilon_r - 1)\mathbf{E} \tag{4.16}$$

from which follows, in combination with the formula $\mathbf{D} = \epsilon_0\epsilon_r\mathbf{E}$,

$$\mathbf{D} = \epsilon_0\mathbf{E} + \mathbf{P} \tag{4.17}$$

In the case of magnetic fields one also encounters three vectors: the flux density **B**, the field intensity **H** and the magnetization **M**; the latter is defined as the magnetic dipole moment of the material per m³. Since we decided in the preceding section to express magnetic dipole moments in ampere m², $M$ must have the dimensions of ampere $m^2\,m^{-3}$ = ampere $m^{-1}$. Hence, $M$ and $H$ have the same dimensions in this system. Note that in the electric case, $P$ has the same dimensions as $D$ rather than as $E$; the reader should thus consider $D$ and $H$ as corresponding quantities, rather than $D$ and $B$, in spite of the similar names of the latter two quantities. Although we do not wish to enter here into a detailed discussion concerning the relationship between the electric and magnetic field vectors, we may point out two reasons for the correspondence between $D$ and $H$, and between $E$ and $B$. One reason lies in the fact that both $E$ and $B$ are defined from force-laws; $E$ from Coulomb's law between two charges, $B$ from the force exerted by a magnetic field on a current [see (4.1)]. The other reason concerns the definitions of $D$ and $H$; $D$ is defined from the theorem of Gauss by a surface integral [see (2.1)], and $H$ is defined in terms of a line integral (see (4.7)). Since $D$ and $H$, and $E$ and $B$ are corresponding quantities, it is not surprising that the formulas encountered in magnetism are *not* analogous to those in dielectrics when considered on the basis of the names of the various quantities; this is somewhat unfortunate, but one has to live with this situation unless one is willing to introduce a completely new nomenclature.

We shall now proceed to derive a relationship between the macroscopic quantities **B**, **H** and **M**, following the line of thought used in section 2.2 for the derivation of expression (4.16). By comparing that section with the discussion below, the reader will discover the correspondence between $D$ and $H$, and between $B$ and $E$ mentioned earlier. Consider a solenoid of length $L$, carrying a current $I$; the total number of turns is $N$. The space inside the solenoid is filled with an isotropic homogeneous material of relative permeability $\mu_r$. We shall assume the solenoid to be ideal in the sense that it produces a homogeneous magnetic field in the material (except

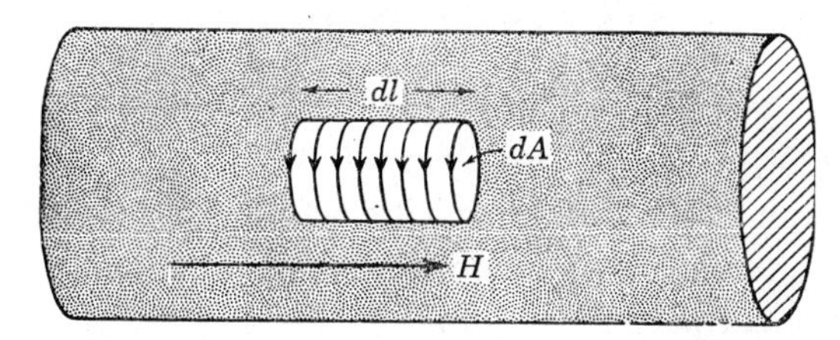


**Fig. 4.6.** Illustrating a cylindrical piece of material subjected to a homogeneous field **H** produced by a solenoid (not drawn). The lines at the surface of the cylindrical cavity of volume $dl\ dA$ represent the current required to keep the flux density inside the cavity equal to the flux density outside the cavity.

near the ends), as indicated in Fig. 4.6. As shown in textbooks on fields, the magnitude of this field is given by

$$H = NI/L \tag{4.18}$$

This formula also follows from (4.7) by choosing an appropriate path of integration. Note that (4.18) does not contain any parameter characteristic of the material inside the solenoid. The flux density in the material is then in accordance with (4.9) given by

$$B = \mu_0\mu_r NI/L \tag{4.19}$$

Suppose now that we cut out of the material a small cylinder with its axis parallel to the original field direction, as indicated in Fig. 4.6. The cross section of the cylinder will be denoted by $dA$, its length by $dl$. How can we achieve a flux density inside the cavity that remains the same as it was when the material was present? Presumably, we are requiring that

$$B_i = B_o = \mu_0\mu_r H \tag{4.20}$$

where the subscript $i$ refers to "inside the cavity" and the subscript $o$

refers to "outside the cavity." Since inside the cavity we have $\mu_r = 1$, the requirement (4.20) may be written in the form

$$\mu_0 H_i = \mu_0 \mu_r H \quad \text{or} \quad H_i - H = (\mu_r - 1)H \tag{4.21}$$

Hence, in order to leave the flux density inside the cavity the same as the flux density outside, the magnetic field $H_i$ inside the cavity must be larger than that outside by an amount $(\mu_r - 1)H$. This can be achieved by letting a current flow along the inside of the cylindrical surface in the same direction as the current in the solenoid, as indicated in Fig. 4.6. How much current is required to produce the extra field $(\mu_r - 1)H$ inside the cavity? Making use of the physical meaning of (4.18), the answer is evidently $(\mu_r - 1)H\, dl$. However, when this current is allowed to flow, the cavity current corresponds to a magnetic dipole moment equal to

$$\mu_m = (\mu_r - 1)H\, dl\, dA \tag{4.22}$$

Since this current serves the same purpose with regard to a uniform flux density as did the material in the cavity before it was taken out, we conclude that in a homogeneous magnetic field, the material carries a magnetic dipole moment per $m^3$ equal to

$$\mathbf{M} = (\mu_r - 1)\mathbf{H} = \chi\mathbf{H} \tag{4.23}$$

where **M** is called the magnetization. This relation between **M** and **H** serves the same purpose in the discussion of magnetic materials as does expression (4.16) in the case of dielectrics. Thus, expression (4.23) forms the *link between the macroscopic theory and the atomic interpretation of the permeability* $\mu_r$. The proportionality constant $\chi$ is called the *magnetic susceptibility* of the material.

The relationship between **B**, **H**, and **M** follows immediately from (4.23); multiplying both sides by $\mu_0$, we find

$$\mu_0\mathbf{M} = \mu_0(\mu_r - 1)\mathbf{H} \quad \text{or} \quad \mathbf{B} = \mu_0(\mathbf{H} + \mathbf{M}) \tag{4.24}$$

The last expression corresponds to (4.17) in the electric case.

## 4.4 Orbital magnetic dipole moment and angular momentum of two simple atomic models

In the preceding sections we have discussed some important concepts pertaining to the macroscopic theory of magnetism. In the present section we shall consider the magnetic dipole moment and its relation to the angular momentum of two simple atomic models. These models are not correct in the sense that they do not represent our present status of knowledge concerning atoms. However, it is useful to consider the properties of these classical models because they exhibit the essential features found in the quantum mechanical interpretation of atoms.

(i) **Circular Bohr orbit.** The first model we shall consider is depicted in Fig. 4.7. It consists of an electron describing a circular orbit of radius $R$ with a stationary nucleus at the center. The charges of the nucleus and

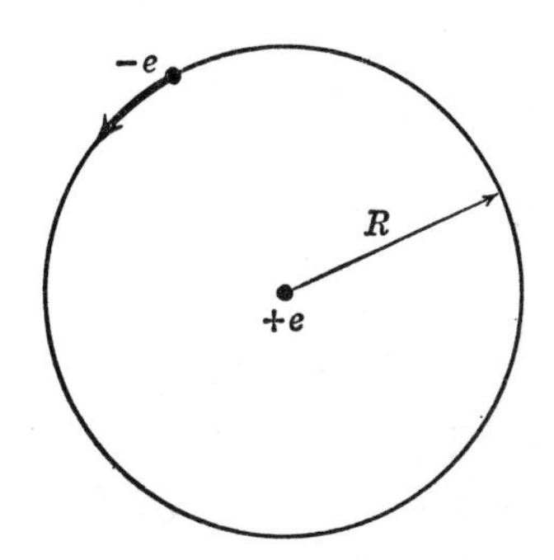


**Fig. 4.7.** Illustrating an electron describing a circular orbit around a proton. The orbital magnetic dipole moment is directed into the paper.

electron are denoted, respectively, by $+e$ and $-e$. We also assume the electron to rotate with a constant angular velocity of $\omega$ radians per second. For the direction of rotation indicated in Fig. 4.7, the motion of the electron in its orbit gives rise to a magnetic dipole moment $\boldsymbol{\mu}_m$ directed into the paper and perpendicular to it. The magnitude of the current associated with the electron motion is evidently equal to $ef$, where $f = \omega/2\pi$ represents the frequency of rotation. Thus, according to (4.14) the magnetic dipole moment of the orbit is

$$|\boldsymbol{\mu}_m| = \pi R^2 e\omega/2\pi = \tfrac{1}{2}e\omega R^2 \tag{4.25}$$

This magnetic dipole moment is called the *orbital magnetic dipole moment*, because it results from the motion of the electron in its orbit around the nucleus. It is of interest to note that there exists a relationship of general validity between the orbital magnetic dipole moment and the *orbital angular momentum.* For the particular case at hand, this relationship may be derived by noting that the angular momentum $\mathbf{M}_a$ is defined as the vector

$$\mathbf{M}_a = m\mathbf{v} \times \mathbf{R} \tag{4.26}$$

where $\mathbf{v}$ is the velocity of the electron and $\mathbf{R}$ the vector which determines its position. Thus, with reference to Fig. 4.7, $\mathbf{M}_a$ is a vector perpendicular to the paper and directed outwardly. Note that $\mathbf{M}_a$ and $\boldsymbol{\mu}_m$ have opposite directions; this is a consequence of the negative charge of the electron. Applying (4.26) to the problem under discussion, we find that

$$M_a = m\omega R^2 \tag{4.27}$$

From (4.25) and (4.27) it then follows that

$$\mu_m = -\frac{e}{2m}\mathbf{M}_a \tag{4.28}$$

Thus, at least for this particular case, we see that the orbital magnetic moment is equal to $(-e/2m)$ times the angular momentum. We shall see later that (4.28) holds for any charge distribution and so may be considered to have general validity for orbital motion of electrons; it is *not* valid for the electron spin or for the nuclear spin, as we shall see in subsequent sections.

From the quantum theory of atoms it follows that the angular momentum of an electron orbit can most conveniently be expressed in units of $h/2\pi$, where $h$ is Planck's constant [$h = 6.62 \times 10^{-34}$ joule sec; note that according to (4.26), $M_a$ has the same dimensions as $h$]. For that reason, one has introduced as an atomic unit of magnetic moment the so-called *Bohr magneton,* defined as

$$1 \text{ Bohr magneton} = \frac{e}{2m}\frac{h}{2\pi} = \frac{eh}{4\pi m}$$

$$= 9.27 \times 10^{-24} \text{ ampere m}^2 \tag{4.29}$$

Since the orbital angular momentum of electrons is of the order of $h/2\pi$, the orbital magnetic moment of an electron in an atom is of the order of 1 Bohr magneton.

**(ii) A spherical charge cloud.** As a second example let us consider an atomic model similar to the one used in section 2.3 in the discussion of the

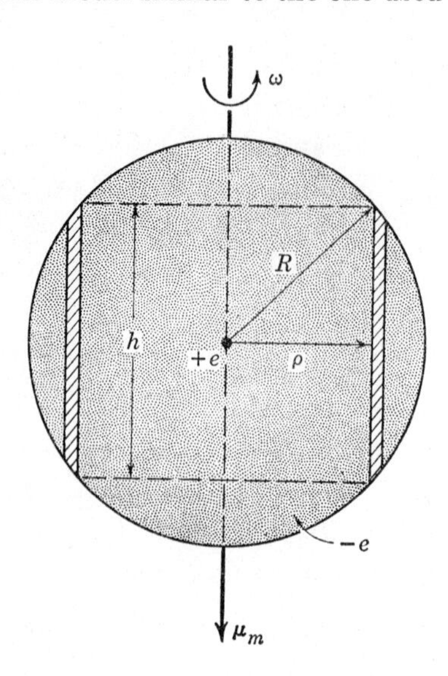


**Fig. 4.8.** Illustrating the calculation of the magnetic dipole moment $\mu_m$ associated with the rotational motion of a charge $-e$ distributed homogeneously throughout a sphere of radius $R$.

polarizability of an atom. In Fig. 4.8 consider a point charge $+e$ surrounded by a negative charge $-e$, the latter being distributed homogeneously throughout a sphere of radius $R$. Assuming that the negative charge cloud rotates with a constant angular frequency $\omega$ around an axis passing through the center of the sphere, what is the magnetic moment of the system? In view of the axial symmetry of the problem we first consider the contribution to the magnetic moment from the charge moving in the cylindrical shell between $\rho$ and $\rho + d\rho$ indicated in Fig. 4.8. The height of this cylinder is evidently equal to

$$h = 2(R^2 - \rho^2)^{1/2} \tag{4.30}$$

The current associated with this shell, i.e. the charge passing per second through a cross section $h\, d\rho$, is equal to

$$di = q\omega\rho h\, d\rho \tag{4.31}$$

where

$$q = -\frac{e}{(4\pi/3)R^3} \tag{4.32}$$

represents the charge density in the cloud. Thus, the contribution to the magnetic dipole moment is

$$d\mu_m = \pi\rho^2\, di = \pi q h \omega \rho^3\, d\rho \tag{4.33}$$

Hence, the total magnetic moment of the system is given by

$$\mu_m = \pi\omega \int_{\rho=0}^{\rho=R} qh\rho^3\, d\rho \tag{4.34}$$

substituting $q$ and $h$ from (4.30) and (4.32), and carrying out the integration one finds

$$\boldsymbol{\mu}_m = -\tfrac{1}{5}eR^2\boldsymbol{\omega} \tag{4.35}$$

The minus sign means that for the configuration given in Fig. 4.8, $\boldsymbol{\mu}_m$ points downwards, as indicated. Comparing the results (4.25) and (4.35) for the two quite different models, it is noted that in both cases the magnitude of the magnetic moment is determined by $e\omega R^2$, and that the results differ only with regard to the numerical constant.

Let us now return for a moment to the expression (4.33), which represents the contribution to $\mu_m$ from the cylindrical shell between $\rho$ and $(\rho + d\rho)$. What is the angular momentum associated with the charge moving in this shell? Applying (4.26) we readily find

$$dM_a = \left[\frac{m}{(4\pi/3)R^3} 2\pi\rho\, h\, d\rho\right] \omega\rho \cdot \rho \tag{4.36}$$

where the term in square brackets represents the mass of the charge between $\rho$ and $(\rho + d\rho)$; we have assumed here that the mass is distributed homogeneously because we had assumed a homogeneous charge distribution. From (4.33) and (4.36), making use of (4.32), it thus follows that

$$d\boldsymbol{\mu}_m = -\frac{e}{2m}\, d\mathbf{M}_a \tag{4.37}$$

This result is identical with that obtained for the circular orbit discussed under (i). Note that the variable $\rho$ does not occur in (4.37) and that the relationship between $d\boldsymbol{\mu}_m$ and $d\mathbf{M}_a$ holds for any volume element of the charge distribution.

## 4.5 Lenz's law and induced dipole moments

In this section we shall pursue the properties of the models discussed in the preceding section somewhat further by investigating the influence of a magnetic field on their behavior. Before doing so, the reader is reminded of the well-known law of Lenz. Thus, in Fig. 4.9(a) consider a loop of wire subjected to a magnetic flux which varies with time. Let $\phi$ be the total flux enclosed by the loop at some instant $t$. Then, if $d\phi/dt$ is not equal to zero, an electric field is set up in the wire, giving rise to an induced current with a direction such that the magnetic field produced by the current counteracts the $d\phi/dt$. Expressed mathematically, this law takes the form

$$\oint \mathbf{E} \cdot d\mathbf{l} = -d\phi/dt \tag{4.38}$$

The line-integral of the electric field along a closed curve is equal to minus the rate of change of the flux enclosed by the curve; the minus sign indicates that the current produced by the electric field counteracts $d\phi/dt$. This law may be applied to any region of space; i.e., the wire loop mentioned only serves the purpose of detecting the existence of an electric field.

It is of interest to realize the difference in behavior between a wire loop and the atomic models to be discussed below with regard to the effect of a varying magnetic flux. Assume, for example, that the flux enclosed by the wire loop varies with time as indicated in Fig. 4.9(b). For $t = 0$, $\phi = 0$; the flux then increases linearly with time until the constant value $\phi_0$ is reached for $t = t_0$. How does the induced current vary with time in this case? According to circuit theory, we may write

$$L\frac{di}{dt} + Ri = -\frac{d\phi}{dt} \tag{4.39}$$

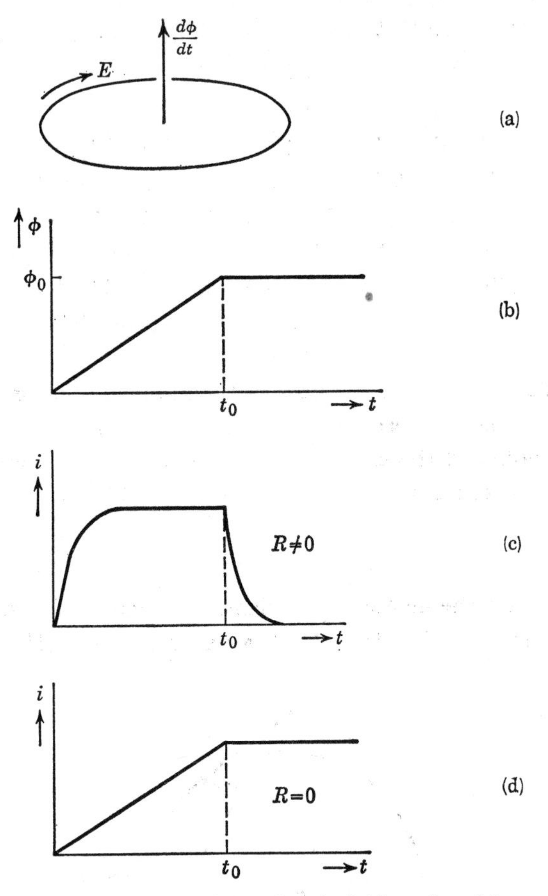


**Fig. 4.9.** The direction of the electric field produced in a wire loop as a result of a change in the enclosed flux is indicated in (a). (b) represents an assumed $\phi(t)$ relationship. (c) and (d) represent the current induced in the wire loop, respectively, for non-zero resistance and for zero resistance.

where $L$ is the self inductance and $R$ the resistance of the loop. In our case, $d\phi/dt$ is constant for the period between 0 and $t_0$, and zero for the period $t > t_0$. The solution of equation (4.39), assuming $t_0 \gg L/R$, will look as indicated schematically in Fig. 4.9(c) for the loop with resistance. Thus, the induced current drops to zero after $t_0$ because there is no longer an electric field. The atomic models to be discussed below are found to behave quite differently. The reason is, that the *electrons in an atom suffer no resistance,* whereas the conduction electrons in a metal wire do. The point we want to make here is that *the atomic models behave as a wire loop with zero resistance.* In fact, if $R$ is zero, (4.39) reduces to

$$L\frac{di}{dt} = -\frac{d\phi}{dt} \tag{4.40}$$

so that for the $\phi(t)$ given in Fig. 4.9(b), a wire without resistance would carry an induced current $i = \phi/L$ as indicated in Fig. 4.9(d). Note that in this case the current remains constant for $t > t_0$. Thus, a permanent change has been accomplished; the current can be made equal to zero only by reducing the flux $\phi$ to zero. We shall now proceed to discuss the influence of a varying magnetic flux on the two atomic models of the preceding section.

(i) **Circular Bohr orbit.** Suppose in the absence of a magnetic field an electron of charge $-e$ describes a circle around a nucleus of charge $+e$; let $R$ be the radius of the orbit, and $\omega_0$ the angular frequency. The orbital magnetic dipole moment in the absence of a field is according to (4.25) equal to

$$\mu_m = -\tfrac{1}{2}eR^2\omega_0 \tag{4.41}$$

Suppose now that the magnetic flux density is increased from zero to some value $B$, where $B$ is directed into the paper in Fig. 4.10. Assuming for

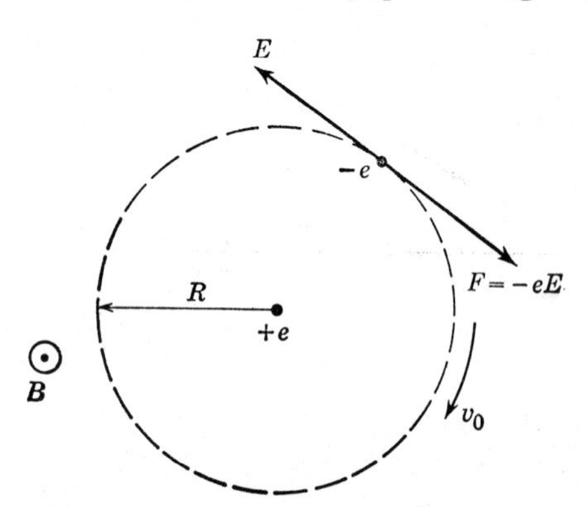


**Fig. 4.10.** The electron describes a circular orbit around a proton with an initial velocity $v_0$ as indicated. A magnetic field of flux density $B$ into the paper is applied. The electric field $E$ and the force $F$ resulting from the change in magnetic flux are indicated.

simplicity that the radius of the orbit remains constant, the electron will experience an electric field $\mathbf{E}$, tangential to the orbit everywhere, equal to

$$E = -\frac{1}{2\pi R}\frac{d\phi}{dt} = -\frac{R}{2}\frac{dB}{dt} \tag{4.42}$$

This follows immediately by applying (4.38) to the present case. The force on the electron during the period that **B** changes with time is then equal to

$$F = -eE = \frac{eR}{2}\frac{dB}{dt} \tag{4.43}$$

The direction of the force **F** for the configuration in Fig. 4.10 is indicated; it is observed that this force tends to accelerate the electron. Now, according to classical mechanics, a force acting during a period $dt$ changes the momentum of a particle in accordance with the equation

$$F\,dt = d(mv) = m\,dv \tag{4.44}$$

where $v$ is the velocity of the particle. In our case, let $\omega(t)$ be the angular frequency of the electron in its orbit at the instant $t$. It then follows from (4.43) and (4.44) that

$$\frac{eR}{2}\frac{dB}{dt}\,dt = mR\,d\omega \quad \text{or} \quad d\omega = \frac{e}{2m}\,dB \tag{4.45}$$

Assuming that for $B = 0$ the angular frequency is $\omega_0$, we find that for any value $B$ the angular velocity is given by

$$\boldsymbol{\omega} = \boldsymbol{\omega}_0 + \frac{e}{2m}\mathbf{B} \equiv \boldsymbol{\omega}_0 + \boldsymbol{\omega}_L \tag{4.46}$$

where $\omega_L$ is called the *Larmor angular frequency.* Since the angular frequency of the electron has changed upon application of the magnetic field, *the orbital magnetic dipole moment has also changed.* In fact, before the field was applied the orbital magnetic moment was

$$\boldsymbol{\mu}_{mo} = -\tfrac{1}{2}eR^2\boldsymbol{\omega}_0$$

and after the field has been applied it is

$$\boldsymbol{\mu}_m = -\frac{1}{2}eR^2\boldsymbol{\omega}_0 - \frac{e^2}{4m}R^2\mathbf{B}$$

The magnetic dipole moment induced by the field is therefore

$$\boldsymbol{\mu}_{m\,\text{ind}} = \boldsymbol{\mu}_m - \boldsymbol{\mu}_{mo} = -\frac{e^2}{4m}R^2\mathbf{B} \tag{4.47}$$

Note that the *induced dipole moment has a direction opposite to the applied magnetic field,* in contrast with the electric dipole moment induced by an electric field (see section 2.3); this result is independent of the initial direction of rotation, as the reader may verify for himself. Also note that since the electron suffers no resistance, it will keep its new angular frequency $\omega$ as long as $B$ remains constant; it thus behaves as the wire loop of zero resistance discussed at the beginning of this section.

An alternative derivation of (4.46) and (4,47) may be given in the following manner: With reference to Fig. 4.11, consider an electron moving in a circular orbit of radius $R$ around the nucleus. In the presence of a magnetic field of flux density **B**, the stability of the orbit requires equilibrium between three forces: (a) the centrifugal force $mv^2/R$; (b) the

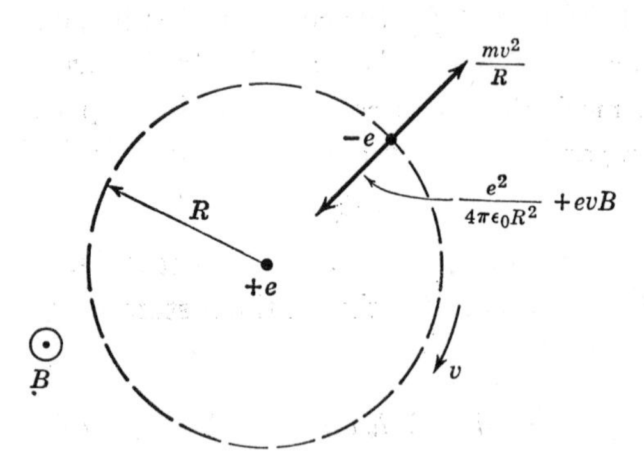


**Fig. 4.11.** Illustrating the equilibrium condition for a circular orbit described by an electron around a proton in the presence of a magnetic field of flux density $B$ into the paper.

Coulomb force $e^2/4\pi\epsilon_0 R^2$ due to attraction by the nucleus; (c) the "Lorentz force" $-e\mathbf{v} \times \mathbf{B}$ due to the magnetic field. For the configuration in Fig. 4.11 we thus require

$$\frac{mv^2}{R} = \frac{1}{4\pi\epsilon_0}\frac{e^2}{R^2} + evB$$

or

$$\omega^2 = \frac{1}{4\pi\epsilon_0}\frac{e^2}{mR^3} + \frac{eB}{m}\omega \tag{4.48}$$

In the absence of a magnetic field, let the angular frequency of rotation be $\omega_0$; then according to (4.48) we obtain by putting $B = 0$

$$\omega_0^2 = \frac{1}{4\pi\epsilon_0}\frac{e^2}{mR^3} \tag{4.49}$$

In the presence of a magnetic field we may therefore write (4.48) in the form

$$\omega^2 = \omega_0^2 + \frac{eB}{m}\omega \tag{4.50}$$

Now, $\omega_0 \cong 10^{15}$ radians per second for the motion of an electron in an atom (see section 3.1). Since the magnetic fields used in the laboratory are of the order of $B \cong 1$ weber m$^{-2}$ or less, we see that $eB/m \cong 10^{11}$ per second which is much smaller than $\omega_0$. Making use of this, one finds readily by solving for $\omega$ from (4.50)

$$\omega \cong \omega_0 + \frac{e}{2m} B \tag{4.51}$$

which is the same as (4.46). The reader may be somewhat astonished by the fact that (4.51) is an approximation whereas it looks as if (4.46) is exact. This is only an apparent contradiction and is a result of the fact that in both derivations we have assumed $R$ to be independent of $B$, which is itself an approximation, valid only as long as $eB/m \ll \omega_0$.

(ii) **Homogeneous spherical charge distribution.** Let us now consider the model consisting of a charge $-e$ distributed homogeneously throughout a sphere of radius $R$; a point charge $+e$ is located at the center of the

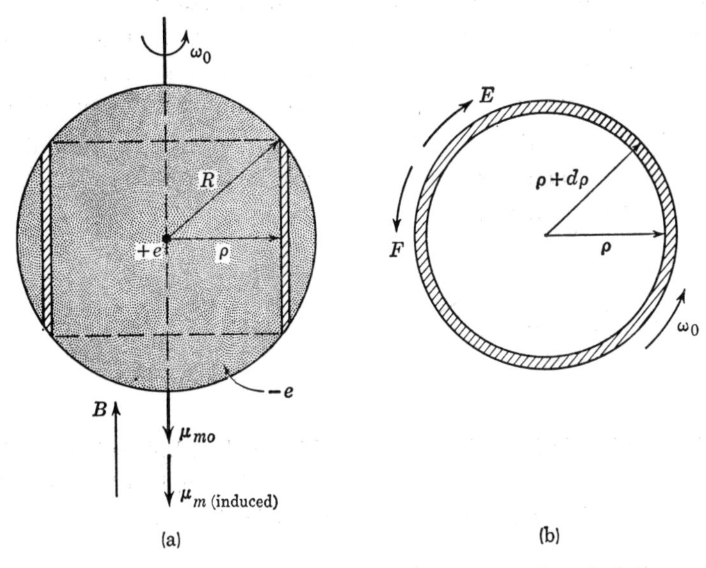


**Fig. 4.12.** Illustrating the geometry pertaining to the calculation of the magnetic dipole moment induced by a homogeneous flux density $B$ in a spherical charge cloud $-e$. In (b) a top view is given, illustrating the cylindrical shell of charge between $\rho$ and $\rho + d\rho$, the electric field $E$ and the force $F$ exerted on the shell of charge.

sphere so as to make the "atom" neutral. In the absence of a magnetic field, let the negative charge cloud rotate around a vertical axis, passing through the center of the sphere, with an angular frequency $\omega_0$ as indicated in Fig. 4.12(a). The magnetic dipole moment of the system is then directed downwards in Fig. 4.12(a), and is given by

$$\boldsymbol{\mu}_{mo} = -\tfrac{1}{5}eR^2\boldsymbol{\omega}_0 \tag{4.52}$$

[see equation (4.35)].

Let us now apply a magnetic field of flux density **B**, where **B** is directed vertically upwards in Fig. 4.12(a). In order to calculate the induced moment in this case we proceed as follows. As in the preceding section, consider the charge rotating in a cylindrical shell between $\rho$ and $\rho + d\rho$ of height $h = 2(R^2 - \rho^2)^{1/2}$. As long as the flux changes, the electric field produced at a distance $\rho$ from the axis is obtained by applying (4.38), giving

$$E = -\frac{1}{2\pi\rho}\frac{d\phi}{dt} = -\frac{\rho}{2}\frac{dB}{dt} \tag{4.53}$$

The field $E$ is tangential to the circle of radius $\rho$ and has a direction as indicated in Fig. 4.12(b). The force exerted by this field on the negative charge thus tends to increase the angular frequency for the configuration assumed in Fig. 4.12(b). The charge in the cylindrical shell divided by the mass in the cylindrical shell is simply equal to $-e/m$, assuming that both charge and mass are distributed homogeneously. From Newton's law (4.44) it thus follows that

$$-\frac{e}{m}E\,dt = \rho\,d\omega \tag{4.54}$$

and since $E(\rho)$ is given by (4.53) we find

$$d\boldsymbol{\omega} = \frac{e}{2m}\,d\mathbf{B} \tag{4.55}$$

Since this result is independent of $\rho$, it holds for the whole sphere of charge. Therefore, if for $B = 0$ the angular frequency is equal to $\omega_0$, we find that for any flux density $B$ the angular frequency is given by

$$\boldsymbol{\omega} = \boldsymbol{\omega}_0 + \frac{e}{2m}\mathbf{B} \equiv \boldsymbol{\omega}_0 + \boldsymbol{\omega}_L \tag{4.56}$$

This result is identical with (4.46) and it will be evident to the reader that the Larmor frequency induced by the magnetic field is independent of the particular charge distribution assumed. The magnitude of the induced dipole moment, of course, does depend on the model. In fact, from (4.35) and (4.56) it follows that for the model under discussion

$$\boldsymbol{\mu}_{m\,\text{ind}} = -\frac{e^2}{10m}R^2\mathbf{B} \tag{4.57}$$

which differs from (4.47) for a circular orbit by a numerical factor. In the derivation given we have assumed tacitly that the charge distribution is independent of the flux density of the applied magnetic field; for practical flux densities obtainable in the laboratory this assumption is justified.

Note that the induced moment is independent of the initial angular frequency $\omega_0$ of the charge distribution. Hence, a magnetic dipole moment given by (4.57) will be induced in the atomic model, independent of whether the model has a "permanent" magnetic dipole moment or not.

# *Part II. Atomic Interpretation of Magnetic Properties of Materials*

## 4.6 Classification of magnetic materials

In this part of the chapter we shall discuss the most essential features of the various types of magnetic materials in terms of the magnetic properties of the atomic dipoles and the interactions between them. The first distinction we can make is that between materials whose atoms carry *permanent magnetic dipoles* and those in which permanent magnetic dipoles are absent; the term permanent magnetic dipole is used here in the same sense as in the corresponding dielectric case; i.e., a permanent dipole exists even in the absence of a field. Materials which lack permanent magnetic dipoles are called *diamagnetic.* If there are permanent magnetic dipoles associated with the atoms in a material, such a material may be *paramagnetic, ferromagnetic, antiferromagnetic,* or *ferrimagnetic,* depending on the interaction between the individual dipoles. Thus, if the interaction between the atomic permanent dipole moments is zero or negligible, a material will be paramagnetic. If the dipoles interact in such a manner that they tend to line up in parallel, the material will be ferromagnetic. If neighboring dipoles tend to line up so that they are antiparallel, the material is antiferromagnetic or ferrimagnetic, depending on the magnitudes of the dipoles on the two "sub lattices," as indicated schematically for a one-dimensional model in Fig. 4.13. Note that in the ferromagnetic case, there is a large resultant magnetization, whereas in an antiferromagnetic configuration the magnetization vanishes. In the case of ferrimagnetic materials, there may be a relatively large net magnetization resulting from the tendency of antiparallel alignment of neighboring dipole moments of unequal magnitude. Ferrimagnetic materials are thus similar to ferromagnetic ones in the sense that both kinds may exhibit a large magnetization. On the other hand, ferrimagnetic materials resemble antiferromagnetic materials with respect to the tendency for antiparallel alignment of neighboring dipole moments.

We should add here a remark to the effect that *induced dipole moments occur in all materials.* In fact, in section 4.5 we showed that a dipole moment induced by a magnetic field in a particular atomic model was inde-

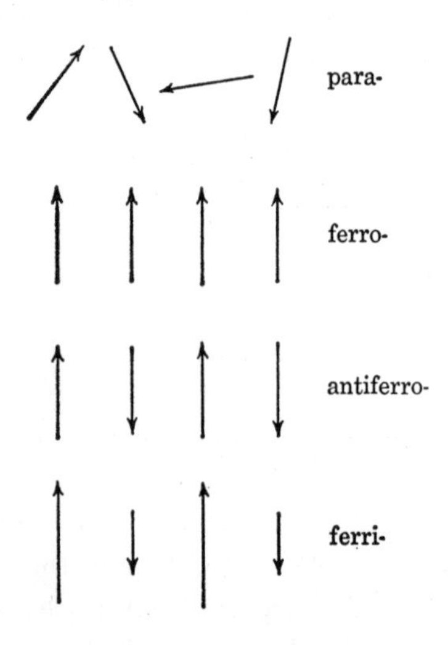


**Fig. 4.13.** Schematic illustration of a paramagnetic, ferromagnetic, antiferromagnetic, and ferrimagnetic arrangement of spins.

pendent of the magnetic dipole moment present in the absence of the field. In this sense then, all materials are diamagnetic. However, when permanent dipole moments are present in numbers comparable to the total number of atoms, the properties of the permanent dipoles usually overshadow the diamagnetic effects and for that reason the classification given above is meaningful.

In the present section we only wish to introduce the classification of magnetic materials; the actual properties of the various kinds of magnetic materials will be discussed in subsequent sections. A summary of the definitions of the various classes of magnetic materials is given in Table 4.1.

**Table 4.1.** Classification of magnetic materials on the basis of the occurrence of permanent atomic magnetic dipoles, and the interaction between them

| Classification | Permanent dipoles | Interaction between neighboring dipoles |
|---|---|---|
| Diamagnetic | No | ——— |
| Paramagnetic | Yes | Negligible |
| Ferromagnetic | Yes | Parallel orientation |
| Antiferromagnetic | Yes | Antiparallel orientation of equal moments |
| Ferrimagnetic | Yes | Antiparallel orientation of unequal moments |

## 4.7 Diamagnetism

The link between the macroscopic and atomic interpretation of magnetism is provided by the formula for the magnetic dipole moment per unit volume, derived in section 4.3,

$$M = (\mu_r - 1)H = \chi H \tag{4.58}$$

It was stated in that section that (4.58) has meaning only if one can define the relative permeability $\mu_r$ for the material under discussion; thus (4.58) is valid for diamagnetic and paramagnetic materials at all temperatures, but for the other classes only above a certain temperature, as we shall see in later sections. The permeability $\mu_r$, or the susceptibility $\chi$ for a diamagnetic or paramagnetic specimen can be determined, for example, by measuring the force exerted on a specimen in an inhomogeneous field (Gouy balance). In Table 4.2 we have given the susceptibility for some

**Table 4.2.** The susceptibility of some diamagnetic materials (at room temperature)

| Material | $\chi = \mu_r - 1$ | Material | $\chi = \mu_r - 1$ |
|---|---|---|---|
| $Al_2O_3$ | $-0.5 \times 10^{-5}$ | Cu | $-0.9 \times 10^{-5}$ |
| $BaCl_2$ | $-2.0 \times 10^{-5}$ | Au | $-3.6 \times 10^{-5}$ |
| NaCl | $-1.2 \times 10^{-5}$ | Ge | $-0.8 \times 10^{-5}$ |
| Diamond | $-2.1 \times 10^{-5}$ | Si | $-0.3 \times 10^{-5}$ |
| Graphite | $-12 \times 10^{-5}$ | Se | $-1.7 \times 10^{-5}$ |

diamagnetic materials. We should note that in the case of metals and semiconductors the susceptibility contains a small paramagnetic contribution associated with the spins of the conduction electrons (the electron spin will be taken up in the next section). It is observed that for these diamagnetic materials the permeability is given approximately by

$$\mu_r \cong 1 - 10^{-5}$$

As long as the electronic structure of the material is independent of temperature, the diamagnetic susceptibility is also essentially independent of temperature. For most engineering applications, $\mu_r$ of a diamagnetic material may be taken as equal to unity.

It is of interest to investigate to what extent the theory of section 4.5 is in agreement with the observed values. From the discussion in section 4.5 it will be evident that an actual calculation of the induced dipole moment would require a detailed knowledge of the electronic structure of the atom. However, an estimate of the order of magnitude of the diamagnetic properties may be obtained by making use of expression (4.57). Assuming that an atom contains, say, 10 electrons, we estimate from (4.57) that the induced magnetic moment should be of the order of

$$\mu_{m\,\text{ind}} \approx -\frac{e^2}{m} R^2 B = -\frac{e^2}{m} R^2 \mu_0 \mu_r H$$

Taking $R \approx 10^{-10}$ m and assuming $N \cong 5 \times 10^{28}$ atoms per m$^3$, we find from $M = N\mu_{m\,\text{ind}} = \chi H$ a value for $\chi$ of the order of $10^{-5}$, in agreement with the experimental values quoted in Table 4.2. There thus seems little doubt that the interpretation of diamagnetism in terms of Lenz's law acting on an atomic scale is essentially correct.

## 4.8 The origin of permanent magnetic dipoles in matter

According to the classification given in section 4.6, the properties of paramagnetic, ferromagnetic, antiferromagnetic, and ferrimagnetic materials are determined by the presence of permanent magnetic dipoles. In this section we shall discuss the various contributions to the permanent magnetic dipole moment of the atomic constituents of matter. According to the results obtained in section 4.4 we can say that whenever a charged particle has an angular momentum, the particle will contribute to the permanent dipole moment. In general, there are three contributions to the angular momentum of an atom:

(i) *orbital angular momentum of the electrons,*
(ii) *electron spin angular momentum,*
(iii) *nuclear spin angular momentum.*

Each of these forms of angular momentum corresponds to a permanent magnetic dipole moment and the total magnetic dipole moment of an atom is obtained by adding the components in an appropriate manner. The rules governing the addition of these components are derived from quantum mechanics and will not be discussed in this book, except in some simple cases. We shall now discuss the contributions separately.

(i) **Orbital magnetic dipole moments.** The relationship between the orbital magnetic dipole moment and the orbital angular momentum has been discussed in terms of a classical model already in section 4.4; we obtained there the relationship [see (4.28)]

$$\boldsymbol{\mu}_m = -\frac{e}{2m} \mathbf{M}_a \qquad (4.28)$$

and this remains valid in the quantum theory. However, quantum theory shows that the orbital angular momentum of an electron in an atom exhibits certain features which are not exhibited by classical models. In section 1.2 we mentioned that the orbital state of motion of an electron in an atom is described by three quantum numbers $n$, $l$ and $m_l$. The *principal quantum number* $n$ determines the energy of the electron; the *orbital quantum number* $l$ determines the orbital angular momentum, and the *magnetic quantum number* $m_l$ determines the component of the angular momentum along an external field direction. The quantum numbers can accept only discrete values, and the rules pertaining to these values as derived from quantum mechanics are the following:

$$\begin{aligned} n &= 1, 2, 3, \ldots \\ l &= 0, 1, \ldots, (n-1) \\ m_l &= l, (l-1), \ldots, 0, -1, \ldots, -l \end{aligned} \tag{4.59}$$

The physical meaning of the magnetic quantum number $m_l$ can be understood within the framework of our present discussion from the following considerations. In atomic physics, angular momentum is measured in units of $h/2\pi$, where $h$ ($= 6.62 \times 10^{-34}$ joule sec) represents Planck's constant. Thus, an electron for which $l = 0$ has no angular momentum and as a consequence of (4.28) also no orbital magnetic dipole moment. An electron for which $l = 1$ can orient itself in such a manner in an applied magnetic field that the components of the angular momentum along the field direction are given by the possible values of $m_l$ as follows:

$$(h/2\pi), \quad 0, \quad -(h/2\pi)$$

These components correspond to the $m_l$ values 1, 0, $-1$, dictated by (4.59). Hence, for $l = 1$, the possible components of the orbital magnetic dipole moment are given by [see (4.28)]

$$-(eh/4\pi m), \quad 0, \quad +(eh/4\pi m) \tag{4.60}$$

as indicated schematically in Fig. 4.14. The reader is reminded here of equation (4.29), which defines the frequently encountered quantity $eh/4\pi m$ as 1 Bohr magneton. In general then, the component of the orbital

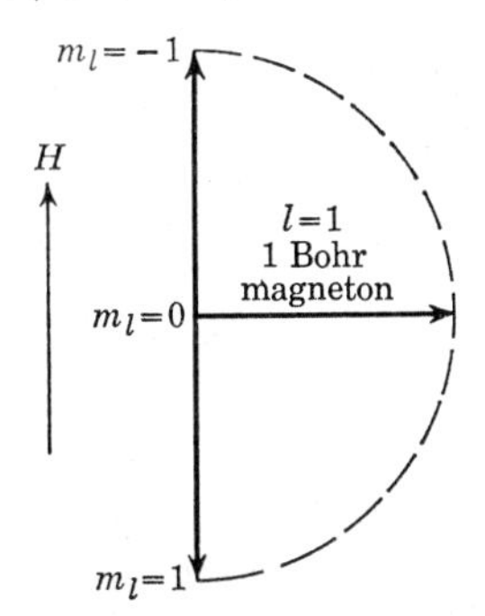


**Fig. 4.14.** Illustrating the three possible components of the magnetic dipole moment in an external field $H$, associated with an orbital momentum quantum number $l = 1$. The total angular momentum is actually equal to $(h/2\pi)\sqrt{l(l+1)}$, and in this sense the figure is somewhat misleading.

magnetic dipole moment along an external field is equal to $-m_l$ Bohr magnetons. From the theory of the periodic table, discussed in Chapter 1, and from the remarks just made, one can readily show that a completely filled electronic shell contributes nothing to the orbital permanent dipole moment of an atom. Consider, for example, the $L$-shell, corresponding to $n = 2$. The possible $l$-values are then 0 and 1. For $l = 0$ there is no magnetic dipole moment anyway. For $l = 1$ we have $m_l = 1, 0, -1$ and if these states are all occupied, the sum of their components vanishes according to (4.60). In general then, one can only expect a resultant orbital magnetic moment in atoms containing incompletely filled electronic shells, and even then the resultant may be zero. Of particular interest to the physicist in this respect are the transition elements; i.e., those elements which have incompletely filled inner shells. A look at Table 1.1, giving the electron configurations of a number of atoms, shows that the elements 21 through 28 (the iron group) fall in this category. Similarly, elements 39 through 45, 58 through 71 (the rare earths) and 89 through 92 are transition elements. For the electrical engineer, the elements of the iron group are of greatest importance. However, in the solid state the orbital magnetic moments of these elements or their compounds are "frozen in." Thus, although the free atoms do have a resultant orbital magnetic moment, the contribution of these moments to the magnetic properties in the solid state is negligible. The reason is that in the iron group the incompletely filled shell lies near the outside of the atoms and is thus highly susceptible to interaction with neighboring atoms in the lattice. As a result of this interaction the dipole moments cannot orient themselves in an external field. In this respect they behave in a way similar to the immobile permanent electric dipole moments in a solid (see Section 2.7).

We should remark here, that for the elements of the rare earths group, the permanent orbital dipole moments do contribute to the magnetic susceptibility. In these elements, the incomplete shells lie relatively deep inside the atom, so that they interact with neighboring atoms to a much smaller degree than do the iron group elements.

In subsequent sections, the contribution from the orbital magnetic dipoles will be neglected, but the reader should realize that this is not always permissible.

(ii) **Electron spin magnetic moment.** In order to explain the details of atomic spectra, Uhlenbeck and Goudmit in 1925 introduced the hypothesis that the electron itself has an angular momentum; i.e., an angular momentum over and above that corresponding to its orbital motion in an atom. The angular momentum of the electron itself is referred to as the *spin* of the electron. Since the electron has a charge, the spin produces a magnetic dipole moment. According to quantum theory, the spin angular momentum along a given direction is either $+h/4\pi$ or $-h/4\pi$; i.e., it can accept only two possible orientations in an external magnetic field. The relationship between the spin angular momentum and the spin magnetic dipole moment is given by

$$\boldsymbol{\mu}_{m\text{ spin}} = -\frac{e}{m}\mathbf{M}_{a\text{ spin}} \tag{4.61}$$

which differs from (4.28) by a factor of 2 on the right-hand side. Thus, the relationship between angular momentum and magnetic dipole moment for the electron spin cannot be understood in terms of a simple classical picture of a rotating sphere of charge. As a result of (4.61) the spin dipole moment components along an external field are

$$+\frac{e}{m}\frac{h}{4\pi} = +1 \text{ Bohr magneton or } -\frac{e}{m}\frac{h}{4\pi} = -1 \text{ Bohr magneton} \tag{4.62}$$

as indicated in Fig. 4.15. In a many-electron atom, the individual spin magnetic moments are added in accordance with certain rules. Here, as in the case of orbital moments, completely filled shells contribute nothing

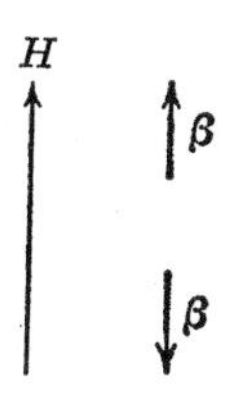


**Fig. 4.15.** Illustrating the two possible moment components associated with the electron spin in an external field $H$; $\beta$ represents 1 Bohr magneton.

to the resultant spin moment. However, an atom such as Na, with one valence electron, has a resultant dipole moment equal to that produced by the valence electron. A sodium ion, $Na^+$, on the other hand, has no resultant spin moment, because the electronic shells are completely occupied. For engineering applications the atoms or ions of the iron group elements are of greatest interest. In Table 4.3 we give the spin configuration associated with the electrons in the incompletely filled $3d$-shell ($n = 3$, $l = 2$) for these elements. The configurations apply to the free atoms as well as to the divalent ions of these elements. In the metallic state, the situation is more complicated and Table 4.3 does not apply. For example, in metallic iron, the average number of Bohr magnetons per atom is 2.2 rather than 4 for the free atom or the $Fe^{2+}$ ion; the non-integral number of Bohr magnetons per atom in the metallic state can be explained in terms of the energy band structure of the transition metals.

**Table 4.3.** NUMBER OF ELECTRONS IN THE $3d$-STATE ($n = 3$, $l = 2$) AND ALIGNMENT OF INDIVIDUAL SPINS FOR THE FREE ATOMS OR DIVALENT IONS OF THE IRON GROUP ELEMENTS; CALCIUM AND COPPER HAVE BEEN ADDED FOR COMPARISON

| Atomic number | Element | $3d$ | Resultant spin in Bohr magnetons |
|---|---|---|---|
| 20 | Calcium | 0 | 0 |
| 21 | Scandium | 1 | 1 ↑ |
| 22 | Titanium | 2 | 2 ↑ ↑ |
| 23 | Vanadium | 3 | 3 ↑ ↑ ↑ |
| 24 | Chromium | 4 | 4 ↑ ↑ ↑ ↑ |
| 25 | Manganese | 5 | 5 ↑ ↑ ↑ ↑ ↑ |
| 26 | Iron | 6 | 4 ↑ ↑ ↑ ↑ ↑ ↓ |
| 27 | Cobalt | 7 | 3 ↑ ↑ ↑ ↑ ↑ ↓ ↓ |
| 28 | Nickel | 8 | 2 ↑ ↑ ↑ ↑ ↑ ↓ ↓ ↓ |
| 29 | Copper | 10 | 0 ↑ ↑ ↑ ↑ ↑ ↓ ↓ ↓ ↓ ↓ |

(iii) **Nuclear magnetic moments.** The angular momentum associated with the nuclear spin is measured in units $h/2\pi$, and is of the same order of magnitude as the electron spin and the orbital angular momentum of the electrons. However, the mass of the nucleus is larger than that of an electron by a factor of the order of $10^3$. Consequently, the magnetic dipole moment associated with the nuclear spin is of the order of $10^{-3}$ Bohr magnetons. Since the nuclear dipole moments are small compared to those associated with the electrons, we may neglect the influence of the former on the magnetic properties of the materials of interest in this book.

In summary then, we shall consider in the following sections only the properties of the electron spin system, assuming that neither the orbital magnetic moments nor the nuclear magnetic moments contribute to the properties of the materials. It should be kept in mind that these omissions are imposed by the limited scope of this book, and that the physicist may be interested, for example, in studying the properties of the nuclear spin system.

## 4.9 Paramagnetic spin systems

In this section we shall consider the susceptibility of a material in as far as it is determined by the presence of electron spin magnetic dipole moments. For simplicity, we shall deal only with a system of spins of one Bohr magneton (such as the scandium atom in Table 4.3); in that case an individual dipole can accept only two possible components along an applied field direction, viz. +1 or −1 Bohr magneton. For atoms with larger spin moments, the calculations are somewhat more complicated, but the essential features are the same. In the present section we shall assume that the interaction between the spins is negligible, so that the field at the position of a given spin may be taken equal to the applied field $H$. This also implies that the flux density at the position of a given spin is assumed to be $B = \mu_0 H$. In making this assumption, we confine ourselves in this section to paramagnetic materials (see the classification in Table 4.1).

Let there be $N$ spins per m³ in the material. In the absence of an applied field, there are as many "up" spins as "down" spins, so that the magnetization $M = 0$. In a field $H$, there will be a preference for the dipoles to line up parallel to the field, and some magnetization will result. At a temperature $T$, let there be $N_p$ dipoles per m³ parallel to the field, and $N_a$ antiparallel; we must then require

$$N_p + N_a = N \tag{4.63}$$

For convenience we shall denote a Bohr magneton by $\beta$, where $\beta = eh/4\pi m$. The magnetization is then given by

$$M = (N_p - N_a)\beta \tag{4.64}$$

Since the macroscopic susceptibility is given by

$$\chi = \mu_r - 1 = M/H \tag{4.65}$$

we wish to express $N_p - N_a$ in terms of $H$, because we shall then be able to express $\chi$ in terms of atomic quantities. As indicated in Fig. 4.16, the

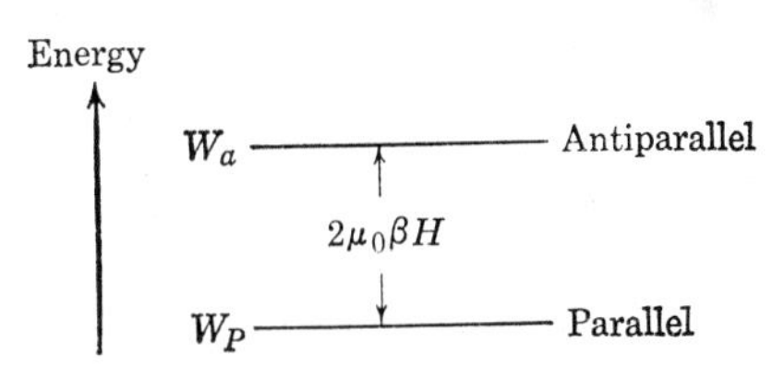


**Fig. 4.16.** Illustrating the energy difference between spin magnetic dipoles parallel and antiparallel to an external field.

energy of a magnetic dipole in the field with antiparallel orientation is larger than that with parallel orientation. The energy difference can be calculated from the fact that the torque, according to (4.15), is in general given by $\boldsymbol{\mu}_m \times \mathbf{B}$, which in our case reduces to $\mu_0 \boldsymbol{\mu}_m \times \mathbf{H}$. It is left to the reader to show that the energy difference between antiparallel and parallel orientation is given by

$$W_a - W_p = 2\mu_0 \beta H \tag{4.66}$$

According to Boltzmann's statistics then, we have for the ratio $N_a/N_p$

the expression

$$N_a/N_p = \exp\,[(W_p - W_a)/kT]$$
$$= \exp\,(-2\mu_0\beta H/kT) \tag{4.67}$$

Thus, we know the sum and the ratio of $N_a$ and $N_p$. It follows from (4.63) and (4.67) that

$$N_p = \frac{N}{1 + \exp\,(-2\mu_0\beta H/kT)} = \frac{N \exp\,(\mu_0\beta H/kT)}{\exp\,(\mu_0\beta H/kT) + \exp\,(-\mu_0\beta H/kT)}$$
$$N_a = \frac{N}{1 + \exp\,(2\mu_0\beta H/kT)} = \frac{N \exp\,(-\mu_0\beta H/kT)}{\exp\,(\mu_0\beta H/kT) + \exp\,(-\mu_0\beta H/kT)} \tag{4.68}$$

Substituting these expressions into (4.64) we find for the magnetization

$$M = N\beta \tanh\,(\mu_0\beta H/kT) \tag{4.69}$$

In Fig. 4.17 we have plotted $M/N\beta$ as a function of the variable $x = \mu_0\beta H/kT$. Note that for $x \ll 1$, $\tanh\,(x) \cong x$, and that for $x \gg 1$,

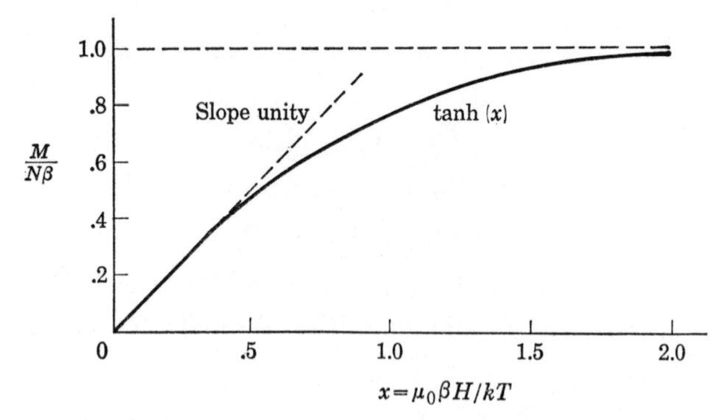


**Fig. 4.17.** The fully drawn curve represents $M/N\beta$ as a function of $x = \mu_0\beta H/kT$. For $x \ll 1$, $\tanh(x) \cong x$, corresponding to a line through the origin of slope unity.

tanh $(x)$ approaches unity. Hence, for strong fields and low temperatures, the magnetization approaches $N\beta$; i.e., it approaches the situation in which all dipoles are lined up in parallel with the field. An example of a paramagnetic salt exhibiting *saturation of the magnetization* is given in Fig. 4.18. For normal temperatures and for not too high fields, $\mu_0\beta H \ll kT$ and under those circumstances $x \ll 1$, so that

$$M \cong N\mu_0\beta^2 H/kT \qquad \text{for} \qquad \mu_0\beta H \ll kT \tag{4.70}$$

In practice, the condition $\mu_0\beta H \ll kT$ is satisfied more often than not. For example, even for a relatively strong field $\mu_0 H \approx 1$ weber m$^{-2}$ we have $\mu_0\beta H \approx 9 \times 10^{-24}$ joule, whereas at room temperature $kT \cong 4 \times 10^{-21}$

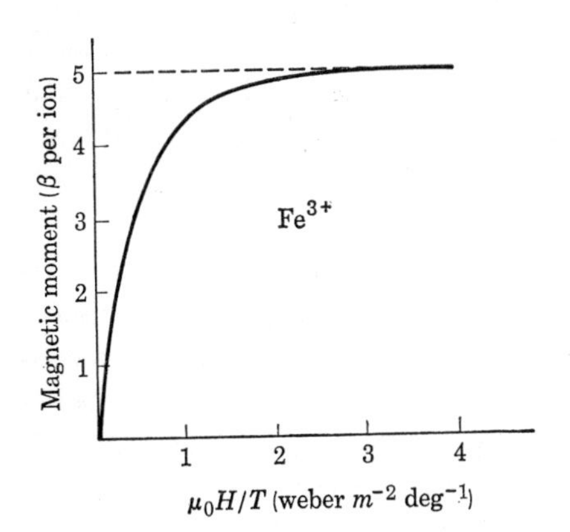


**Fig. 4.18.** The magnetic moment in Bohr magnetons per $Fe^{3+}$ ion in ferric ammonium alum as a function of $\mu_0 H/T$; note the observed approach to the saturation value of $5\beta$. [After W. E. Henry, *Phys. Rev.* **88**, 559 (1952)]

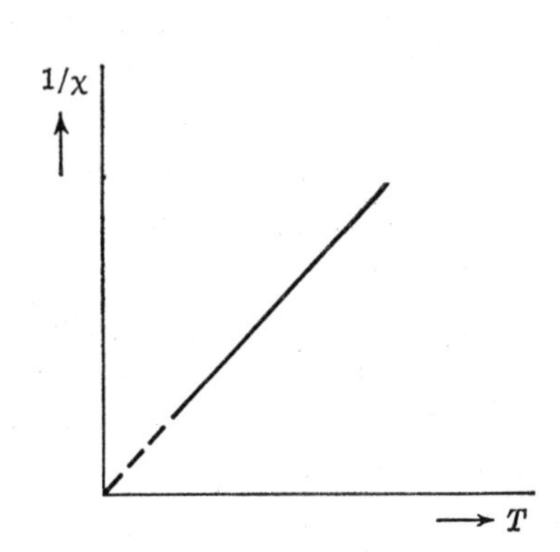


**Fig. 4.19.** The reciprocal of the susceptibility as a function of $T$ for a paramagnetic material, illustrating the Curie law.

joule. Assuming that (4.70) is valid, we find from (4.65) and (4.70) for the susceptibility

$$\chi = \mu_r - 1 = N\mu_0\beta^2/kT \equiv C/T \tag{4.71}$$

Thus, the susceptibility varies as $1/T$; it is instructive in this connection to compare the similar problem of orientational polarization in dielectrics, discussed in section 2.5. The law expressed by (4.71) is known as the *Curie law* of paramagnetism; it is illustrated in Fig. 4.19. The constant $C = N\mu_0\beta^2/k$ is called the *Curie constant*.

An estimate of the magnitude of $\chi$ (or $\mu_r$) at a given temperature may be made by taking $N \cong 5 \times 10^{28}$ m$^{-3}$. Putting in numerical values for the other quantities in (4.71), one finds $\chi \cong 0.3/T$; i.e., $\chi$ is of the order of $10^{-3}$ at room temperature. Experimentally determined values are given in Table 4.4; it is observed that these are of the estimated order of magni-

**Table 4.4.** Susceptibilities of some paramagnetic materials at room temperature

| Substance | $\chi = \mu_r - 1$ | Substance | $\chi = \mu_r - 1$ |
|---|---|---|---|
| $CrCl_3$ | $1.5 \times 10^{-3}$ | $Fe_2O_3$ | $1.4 \times 10^{-3}$ |
| $Cr_2O_3$ | $1.7 \times 10^{-3}$ | $Fe_2(SO_4)_3$ | $2.2 \times 10^{-3}$ |
| CoO | $5.8 \times 10^{-3}$ | $FeCl_2$ | $3.7 \times 10^{-3}$ |
| $CoSO_4 \cdot H_2O$ | $2.0 \times 10^{-3}$ | $FeSO_4$ | $2.8 \times 10^{-3}$ |
| $MnSO_4$ | $3.6 \times 10^{-3}$ | $NiSO_4$ | $1.2 \times 10^{-3}$ |

tude. It should be realized that the measured susceptibility includes a diamagnetic contribution which has not been considered in the present section. However, since susceptibilities are additive quantities, and since $\chi_{\text{dia}} \approx 10^{-5}$ according to the results in section 4.7, we see that $\chi_{\text{dia}} \ll \chi_{\text{para}}$ at room temperature and below.

For many applications in electrical engineering, it is a good approximation to take the relative permeability $\mu_r$ of paramagnetic substances equal to unity. As far as applications of paramagnetic materials are concerned, we may mention here that paramagnetic salts are the working material used in obtaining very low temperatures ($< 1°$K) by *adiabatic demagnetization:* the principle of this method is discussed in the books by Kittel and by Dekker, given in the list of general references. Also, paramagnetic salts have entered the group of electrical engineering materials a few years ago because they are the essential material used in the solid state *maser* (microwave amplification through stimulated emission by radiation). The principle of operation of a maser is discussed in van der Ziel's book (page 590 ff), cited in the general references.

## 4.10 Some properties of ferromagnetic materials

Each ferromagnetic material has a characteristic temperature above which its properties are quite different from those below that temperature. This temperature is called the *ferromagnetic Curie temperature* and will be denoted here by $\theta_f$. In this section we shall discuss briefly some of the characteristic features of ferromagnetic behavior in the two temperature regions.

(i) $T > \theta_f$. In the region above the ferromagnetic Curie temperature, the behavior of a ferromagnetic material is somewhat similar to that of a paramagnetic material. Thus, there exists a unique relationship between $B$ and $H$, and between $M$ and $H$. One can thus define the susceptibility $\chi = M/H = \mu_r - 1$, where $\chi$ and $\mu_r$ have a definite meaning. In this region, the susceptibility depends on temperature in accordance with the so-called *Curie-Weiss law*

$$\chi = \mu_r - 1 = C/(T - \theta) \qquad \text{for} \qquad T > \theta_f \tag{4.72}$$

$C$ is called the *Curie constant;* $\theta$ is the "*paramagnetic*" *Curie temperature.* This expression is not valid in the region close to $\theta_f$, as may be seen from Fig. 4.20; note that in this figure $1/\chi$ is plotted as a function of $T$. Comparison of Fig. 4.20 and Fig. 4.19 shows that the ferromagnetic case and the paramagnetic case are very similar; the only difference is that for a

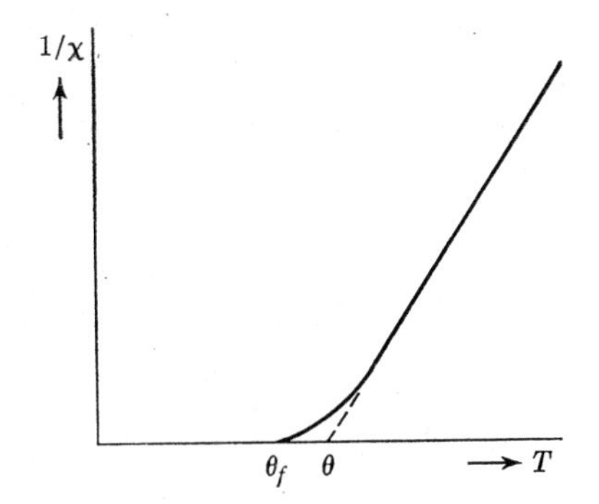


**Fig. 4.20.** The reciprocal of the susceptibility as a function of temperature for a ferromagnetic material above the ferromagnetic Curie temperature, $\theta_f$. The paramagnetic Curie temperature $\theta$ is obtained by extrapolation of the straight portion of the curve which satisfies the Curie-Weiss law.

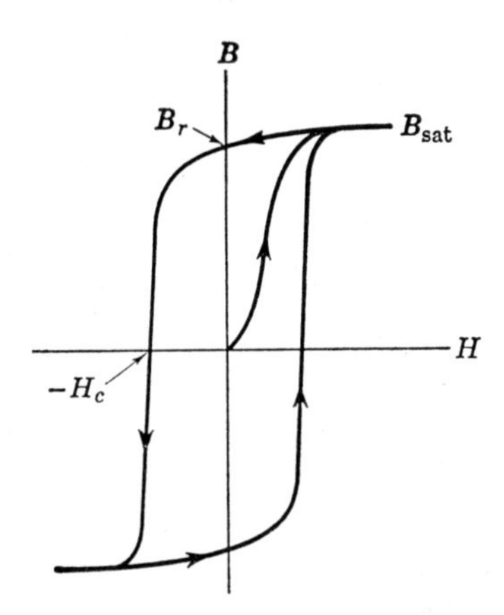


**Fig. 4.21.** Schematic representation of the hysteresis loop for a ferromagnetic material. The virgin curve starts at the origin.

truly paramagnetic material $\theta = 0$. The paramagnetic Curie temperature $\theta$ is usually somewhat higher than the ferromagnetic Curie temperature (see Fig. 4.20); for the ferromagnetic elements of the iron group, for example, these temperatures are, in degrees absolute:

| | Fe | Co | Ni |
|---|---|---|---|
| $\theta_f$ . . . | 1043 | 1393 | 631 |
| $\theta$ . . . | 1093 | 1428 | 650 |

(ii) $T < \theta_f$. Below the ferromagnetic Curie temperature, ferromagnetic materials exhibit the well-known *hysteresis* in the $B$ versus $H$ curves. A schematic representation of this behavior is given in Fig. 4.21. Starting with a virgin specimen, $B$ varies reversibly with $H$ for small fields. Since there is no hysteresis in this region, one defines the *"initial" permeability* $\mu_r$ in the same way as the permeability of a paramagnetic material. As the field $H$ is increased, $B$ begins to increase rapidly and ultimately approaches a saturation value $B_{sat}$. Along the virgin curve, one can still speak of a *differential permeability* defined by $1 + (dM/dH)$, but evidently this value is a function of $H$ itself. The differential permeability may become very large, as is evident from the values given in Table 4.5 for high-permeability materials. Upon reducing the value of $H$ from the saturation region to zero, it is observed that there remains a flux density $B_r$ (*remanent flux density*). Since $H = 0$, the material must be spontaneously magnetized; in fact, the magnetization corresponding to $B_r$ is equal to $M_r = B_r/\mu_0$. The occurrence of *spontaneous magnetization* is characteristic of ferromagnetic materials; in this respect they behave in a way similar to ferroelectrics.

The field $-H_c$ required to reduce the flux density to zero is called the *coercive force*. The coercive force of ferromagnetic materials varies over a wide range of values. For example, the coercive force of supermalloy, used in pulse transformers, is approximately 1 ampere $m^{-1}$, whereas that for a high stability permanent magnet may be as high as $10^6$ ampere $m^{-1}$. The coercive force thus determines to a large extent the practical applications for which a given material may be used. Some data referring to the magnetic properties of a number of ferromagnetic materials are given in Table 4.5.

**Table 4.5.** Some data pertaining to ferromagnetic materials ($B_{sat}$ is the saturation flux density; $B_r$ is the remanent flux density; $H_c$ is the coercive force and $(\mu_r)_{max}$ is the maximum differential permeability.)

| High permeability materials | $(\mu_r)_{max}$ | $B_{sat}$ (weber $m^{-2}$) | $H_c$ (amp $m^{-1}$) |
|---|---|---|---|
| Iron | 5000 | 2.1 | 80 |
| 4% Si-Fe | 7000 | 2.0 | 40 |
| Mu metal | $10^5$ | 0.65 | 4 |
| Supermalloy | $8 \times 10^5$ | 0.8 | 0.16 |

| Permanent magnet materials | $B_r$ (weber $m^{-2}$) | $H_c$ (amp $m^{-1}$) |
|---|---|---|
| Carbon steel | 1 | 4000 |
| Alnico V | 1.25 | 44,000 |
| Platinum-Cobalt | 0.45 | $2 \times 10^5$ |

## 4.11 Spontaneous magnetization and the Curie-Weiss law

In this section we shall discuss the atomic interpretations of spontaneous magnetization and of the Curie-Weiss law. Before doing this, we wish to point out that a piece of valuable information regarding the interpretation of ferromagnetic behavior may be gained by considering the magnitude of the remanent flux density of permanent magnets. We see from Table 4.5 that $B_r \approx 1$ weber $m^{-2}$ for these materials, and since $H = 0$ we conclude that the remanent magnetization $M_r = B_r/\mu_0 \approx 10^6$ ampere $m^{-1}$. On the other hand, we know that an atomic dipole is of the order of 1 Bohr magneton, i.e. $\cong 10^{-23}$ ampere $m^2$. We thus require approximately $10^{29}$ atomic dipoles per m$^3$, all lined up in parallel, to obtain the observed magnetization. However, the number of atoms in a solid is approximately $10^{29}$ per m$^3$, so that the observed $M_r$ indicates *parallel alignment of essentially all the dipoles in the material.* This notion brings us to the first hypothesis of Weiss, who by 1907 had already suggested that in ferromagnetic materials the internal field seen by a given dipole is equal to the applied field plus a contribution from the neighboring dipoles which tends to align it in the same direction as its neighbors. Weiss expressed this mathematically by stating that the internal field $\mathbf{H}_i$ is given by*

$$\mathbf{H}_i = \mathbf{H} + \gamma\mathbf{M} \tag{4.73}$$

$\mathbf{H}$ is the applied field and $\gamma\mathbf{M}$ is a measure for the tendency of the environment to align a given dipole parallel to the magnetization already existing. The proportionality constant $\gamma$ is the *internal field constant;* it determines the strength of the interaction between the dipoles (see the classification of magnetic materials in Table 4.1). We shall now show that a field of the type (4.73) is consistent with (a) the Curie-Weiss law and, (b) the occurrence of spontaneous magnetization. As a model we shall again consider a system of $N$ spins per m$^3$, each giving rise to a magnetic moment of 1 Bohr magneton, $\beta$, either parallel or antiparallel to an external field. The magnetization of such a system may be obtained immediately from expression (4.69) for the paramagnetic case, by replacing $H$ by $H_i$. Hence,

$$M = N\beta \tanh\left[\frac{\mu_0\beta}{kT}(H + \gamma M)\right] \tag{4.74}$$

At this point it is convenient to distinguish between two temperature regions:

(i) **High temperatures.** At sufficiently high temperatures, the term in square brackets in (4.74) will become small compared to unity. Then, since $\tanh x \cong x$ for $x \ll 1$, we may approximate (4.74) by

$$M = (N\mu_0\beta^2/kT)(H + \gamma M) \tag{4.75}$$

Solving this equation for $M$, one finds for the susceptibility of the material

$$\chi = \frac{M}{H} = \frac{N\mu_0\beta^2/k}{T - N\mu_0\beta^2\gamma/k} = \frac{C}{T - \theta} \tag{4.76}$$

Note that this expression is identical in form with the Curie-Weiss law (4.72). For the model studied here, we have

$$C = N\mu_0\beta^2/k \qquad \text{and} \qquad \theta = \gamma C \tag{4.77}$$

* Compare expression (2.32) for the case of dielectrics.

Since $C$ and $\theta$ can be determined from measurements of the susceptibility as a function of temperature, the internal field constant $\gamma$ can be calculated. One finds for ferromagnetic materials $\gamma \approx 10^3$. This value is about a thousand times as large as one would obtain on the assumption that the internal field is due to the *magnetic* interaction of the atomic dipoles (see problem 4.11). In fact, the forces acting between the dipoles in a ferromagnetic material cannot be explained in terms of classical physics; they are due to the wave nature of the electrons and in wave mechanics are called *exchange forces*.

(ii) **Spontaneous magnetization below the Curie temperature.** It follows from (4.76) that the Curie-Weiss law can hold only for temperatures $T > \theta$, because for $T = \theta$ the susceptibility would become infinite. This fact suggests already that at $T = \theta$, spontaneous magnetization may occur (non-vanishing $M$ for $H = 0$); this is confirmed by the following arguments. In (4.74) let us put $H = 0$, and ask the question as to whether that equation permits a non-vanishing value for $M$. It is convenient to introduce a new variable

$$x = \gamma\mu_0\beta M/kT \tag{4.78}$$

so that (4.74) may be written (with $H = 0$) in the form

$$M/N\beta = M/M_{\text{sat}} = \tanh x \tag{4.79}$$

Here, $M_{\text{sat}} = N\beta$ represents the saturation value of the magnetization, since it gives the magnetization for parallel alignment of all the dipoles, and is evidently the maximum value that can be obtained. A plot of $M/M_{\text{sat}}$

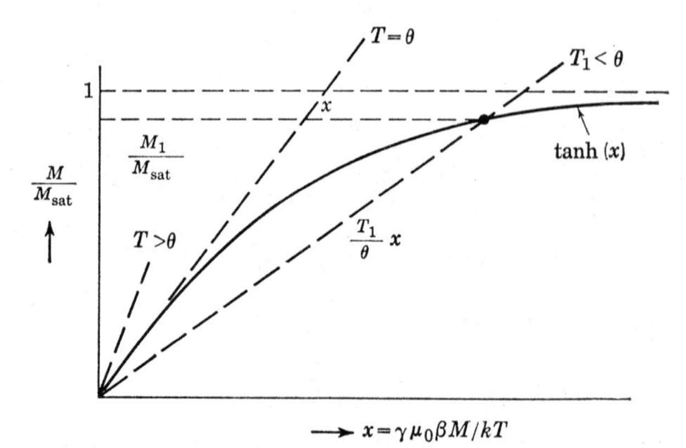


**Fig. 4.22.** Schematic illustration of the graphical solution of equation (4.79) for the spontaneous magnetization. For the temperature $T_1(< \theta)$ the value of $M_1/M_{\text{sat}}$ is obtained from the intersection of the line $(T_1/\theta)x$ and the $\tanh(x)$ curve. For $T \gg \theta$ the spontaneous magnetization vanishes.

as function of $x$ is given in Fig. 4.22. According to (4.78), we should also have

$$\frac{M}{M_{\text{sat}}} = \frac{M}{N\beta} = \frac{kT}{N\gamma\mu_0\beta^2}\,x = \frac{T}{\theta}\,x \tag{4.80}$$

where the last equality follows from (4.77). Now, for a given temperature $T$, (4.80) in a plot of $M/M_{\text{sat}}$ versus $x$ represents a straight line with a slope equal to $T/\theta$. Since $M/M_{\text{sat}}$ must satisfy both (4.79) and (4.80), the value of $M/M_{\text{sat}}$ for the temperature $T$ is given by the intersection of the straight line and the tanh $x$ curve, as indicated in Fig. 4.22. When this procedure is repeated for different temperatures, one can finally plot $M/M_{\text{sat}}$ as function of $T/\theta$, as shown in Fig. 4.23. Note that for $T \geqslant \theta$, the spon-

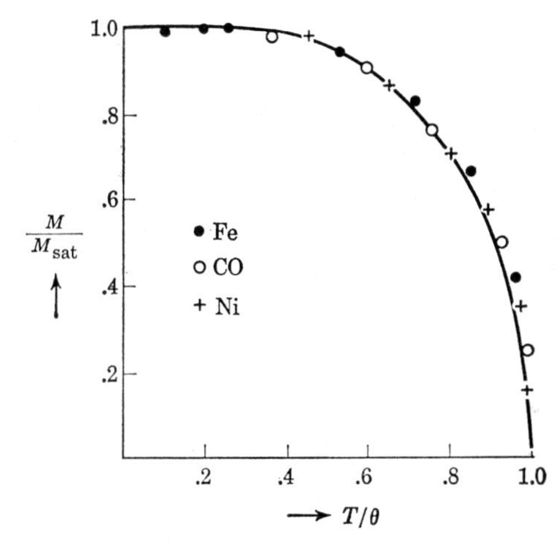


**Fig. 4.23.** The curve represents the relative spontaneous magnetization as a function of $T/\theta$ obtained from the procedure illustrated in Fig. 4.22. The points represent measured values for nickel, cobalt, and iron.

taneous magnetization vanishes. The reason for this can be seen in Fig. 4.22. When $T = \theta$, expression (4.80) gives $M/M_{\text{sat}} = x$, but this line is just the tangent of the tanh $x$ curve at the point $x = 0$. For $T \geqslant \theta$, the only intersection between (4.80) and (4.79) is the point $x = 0$; i.e., there is no longer a spontaneous magnetization. In Fig. 4.23, experimental points have been indicated, and the agreement with the theoretical curve is seen to be quite good. The reader should realize that the materials for which the experimental points have been plotted have widely different $\theta$ and $M_{\text{sat}}$ values, as indicated in Table 4.6. The spontaneous magnetization becomes equal to $M_{\text{sat}}$ only at $T = 0$, but it is evident from the curve in

**Table 4.6.** Saturation value of the spontaneous magnetization and ferromagnetic Curie temperatures for the ferromagnetic metals

| Metal | $M_{sat}$ (amp m$^{-1}$) | Curie temp. (°K) |
|---|---|---|
| Fe | $1.75 \times 10^6$ | 1043 |
| Co | $1.45 \times 10^6$ | 1393 |
| Ni | $0.51 \times 10^6$ | 631 |

Fig. 4.23 that for iron and cobalt even at room temperature, the spontaneous magnetization is nearly equal to $M_{sat}$.

It is noted that the theory given predicts the same value for the ferromagnetic Curie temperature as for the paramagnetic Curie temperature, whereas experimental values of $\theta_f$ and $\theta$ differed somewhat. This discrepancy between theory and experiment must be ascribed to the simple form of the internal field expressed by (4.73). On the other hand, it must be admitted that this simple equation explains the spontaneous magnetization and the Curie law satisfactorily as far as the essential features are concerned.

## 4.12 Ferromagnetic domains and coercive force

After the discussion in the preceding section, the question may be raised as to how one can explain the fact that a piece of iron may not exhibit a resultant magnetization, and how one can explain the hysteresis in the $B$ versus $H$ curves of ferromagnetic materials. This brings us to the *second hypothesis of Weiss.* According to Weiss, a virgin specimen of iron consists of a number of regions or *domains* ($\approx 10^{-6}$ m or larger) which are spontaneously magnetized in accordance with the formulas derived in the preceding section. However, the *direction of the spontaneous magnetization varies from domain to domain,* and consequently, the resultant magnetization may be zero or nearly zero; this is indicated by the domain configuration in Fig. 4.24(a). If an external field $H$ is applied, the domains with the proper direction of spontaneous magnetization grow at the expense of those that are magnetized in other directions by virtue of a motion of the domain walls [see Fig. 4.24(b)]. Ultimately, as the field is increased, the whole specimen may become one single domain, and saturation has been achieved. Thus, the hysteresis curve is associated with the motion of domain walls and, to some extent by domain rotation. The latter takes place as a result of the fact that spontaneous magnetization occurs only along certain directions in the crystal; when a field is applied in another arbitrary direction, the magnetization will rotate from an "easy"

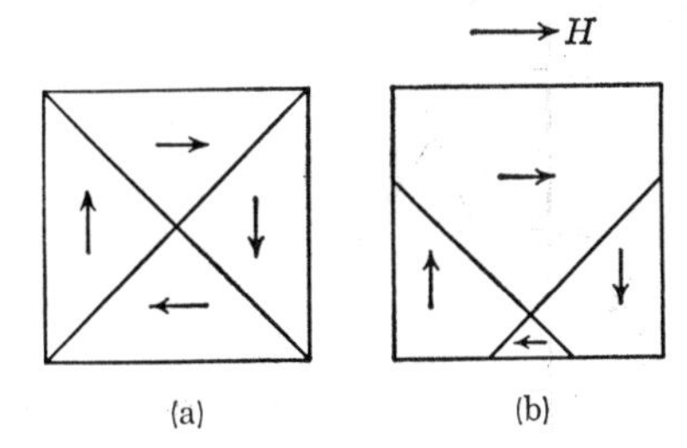


**Fig. 4.24.** The domain configuration in (a) has zero resultant magnetization. In (b) a magnetic field has been applied and the domain walls have moved so as to produce a net magnetization along the applied field direction.

direction to a "hard" direction. Since the hysteresis loop is interpreted in terms of domain wall motion, the coercive force must be determined by the "mobility" of the walls. The mobility of the domain walls is in turn determined by impurities, lattice imperfections, etc. and to some extent it is possible to "design" materials which require a large or a small coercive force. It may be mentioned here that the well-known Barkhausen effect is due to irregular fluctuations in the motion of domain walls; in earlier days, the effect had been ascribed to rotation of domains.

The most direct evidence for the existence of domains is provided by the so-called *Bitter powder patterns*. A drop of a colloidal suspension of ferromagnetic particles is placed on a well-prepared surface of the specimen; since there are strong local magnetic fields near the domain boundaries, the particles congregate there and the domain structure may be observed under a microscope.

## 4.13 Antiferromagnetic materials

In the discussion of ferromagnetic materials it was pointed out that the tendency for parallel alignment of the electron spins was due to quantum mechanical exchange forces. In certain materials, for example when the distance between the interacting atoms is small, the exchange forces produce a tendency for *antiparallel alignment* of electron spins of neighboring atoms. This kind of interaction is encountered in antiferromagnetic and in ferrimagnetic materials. It is of interest to note that certain properties of antiferromagnetic materials were predicted before antiferromagnetism was discovered experimentally. Thus, Néel and Bitter in the thirties made a theoretical study of the properties of antiferromagnetic models, and a few years later, in 1938, antiferromagnetism was discovered in MnO by Bizette, Squire and Tsai. Since that time, a number of other materials has been found to be antiferromagnetic. From the experimental point of view, the most characteristic feature of an antiferromagnetic material is the occurrence of a rather sharp maximum

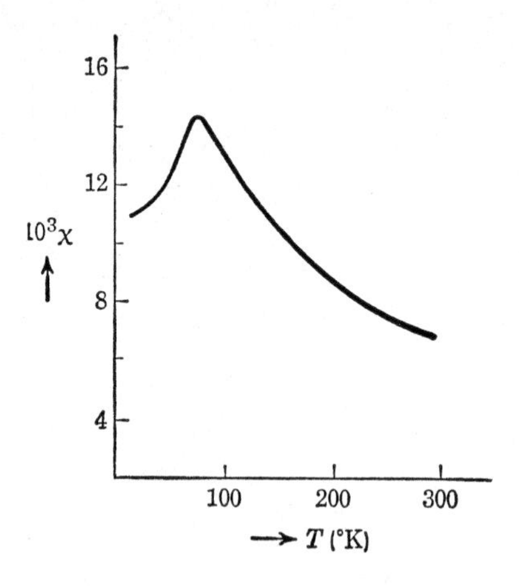


**Fig. 4.25.** The susceptibility of $MnF_2$ (polycrystalline) as a function of temperature. The maximum is characteristic of an antiferromagnetic transition. [After de Haas, Schultz, and Koolhaas, *Physica*, **7**, 57 (1940)]

in the susceptibility-versus-temperature curve, as may be seen from the example given in Fig. 4.25 for $MnF_2$. The temperature for which this maximum occurs is called the *Néel temperature*, $T_N$. The Néel temperature plays a similar role in antiferromagnetic materials as does the ferromagnetic Curie temperature in ferromagnetic materials. Thus, above the Néel temperature, the susceptibility is observed to follow the equation

$$\chi = \frac{C}{T + \theta} \tag{4.81}$$

where $C$ is the Curie constant and $\theta$ the paramagnetic Curie temperature. Below the Néel temperature, the spin system tends to be "ordered" in a way similar to the spin system in a ferromagnetic material, except that at $T = 0$ half the spins are oriented in one direction and the other half in the opposite direction. Confining ourselves for the moment to the high temperature region, it is of interest to recapitulate the results for the susceptibility versus temperature behavior for para-, ferro-, and antiferromagnetic materials:

$$\begin{array}{ccc} \textit{para-} & \textit{ferro-} & \textit{antiferro-} \\ \chi = C/T & \chi = C/(T - \theta) & \chi = C/(T + \theta) \\ & \text{for } T > \theta_f & \text{for } T > T_N \end{array} \tag{4.82}$$

The difference between the three groups of materials is illustrated in Fig. 4.26 in terms of a plot of $1/\chi$ versus temperature.

We shall now consider a simple model of an antiferromagnetic material. With reference to Fig. 4.27, consider a body centered cubic lattice. We shall distinguish in this lattice between $A$-sites and $B$-sites, as indicated in Fig. 4.27. Each $A$-site is surrounded by eight $B$-sites, and each

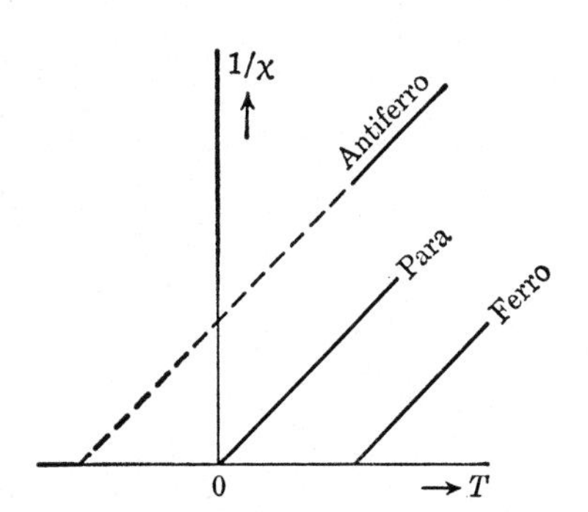


Fig. 4.26. Illustrating the reciprocal susceptibility as a function of temperature for a paramagnetic, ferromagnetic, and antiferromagnetic material.

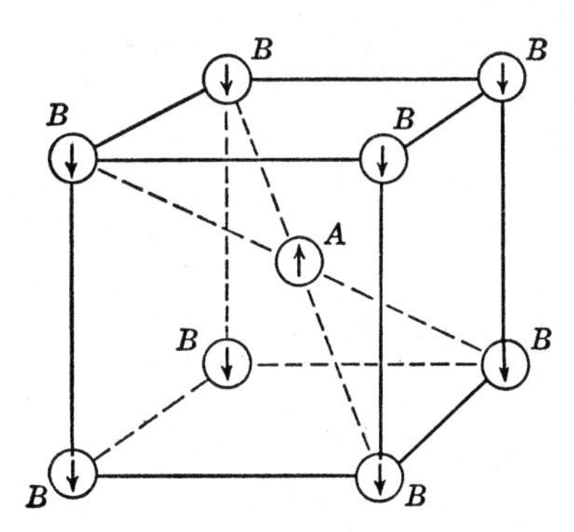


Fig. 4.27. Representation of two sub-lattices, $A$ and $B$; the spins on the $A$-lattice tend to line up antiparallel to those on the $B$-lattice.

$B$-site is surrounded by eight $A$-sites. We shall assume that all sites are occupied by identical atoms with a magnetic dipole moment of 1 Bohr magneton which can orient itself either in the "up" or the "down" direction. Also, we shall assume that an atom at an $A$-site tends to align its spin opposite to the spins of the neighboring atoms on the $B$-sites, and vice versa. In order to describe this mathematically, we introduce an internal field $\mathbf{H}_a$ for the atoms on $A$-sites and an internal field $\mathbf{H}_b$ for the atoms on $B$-sites. Following a procedure similar to that used in section 4.11 for a ferromagnetic material, we may then write

$$\mathbf{H}_a = \mathbf{H} - \gamma \mathbf{M}_b \quad \text{and} \quad \mathbf{H}_b = \mathbf{H} - \gamma \mathbf{M}_a \tag{4.83}$$

Here, $\mathbf{M}_a$ and $\mathbf{M}_b$ represent the magnetizations of the $A$-sub-lattice and of the $B$-sub-lattice respectively. The minus signs represent the assumption that an $A$ atom tends to align its dipole moment opposite to the direction of the magnetization of the $B$-lattice; the internal field constant $\gamma$ determines the strength of the exchange interaction. The magnetizations $M_a$ and $M_b$ may be obtained again from formula (4.69) by replacing $H$ by the appropriate internal fields. Hence, if there are $N$ atoms per $m^3$ on the $A$-lattice and an equal number of atoms on the $B$-lattice, we find

$$\begin{aligned} M_a &= N\beta \tanh\left[\frac{\mu_0\beta}{kT}(H - \gamma M_b)\right] \\ M_b &= N\beta \tanh\left[\frac{\mu_0\beta}{kT}(H - \gamma M_a)\right] \end{aligned} \tag{4.84}$$

At sufficiently high temperatures, we may make use of the fact that tanh $\cong x$ for $x \ll 1$, so that then (4.84) reduces to

$$\mathbf{M}_a = (N\beta^2\mu_0/kT)(\mathbf{H} - \gamma\mathbf{M}_b)$$
$$\mathbf{M}_b = (N\beta^2\mu_0/kT)(\mathbf{H} - \gamma\mathbf{M}_a) \tag{4.85}$$

The total magnetization of the material is given by

$$\mathbf{M} = \mathbf{M}_a + \mathbf{M}_b \tag{4.86}$$

and consequently may be calculated by adding the two (4.85) equations; this gives

$$\mathbf{M} = (N\mu_0\beta^2/kT)(2\mathbf{H} - \gamma\mathbf{M}) \tag{4.87}$$

Since the net magnetization **M** must have the same direction as **H**, we may consider (4.87) as a scalar equation. Solving for $M/H$ we thus find for the susceptibility at high temperatures

$$\chi = M/H = 2C/(T + \gamma C) = 2C/(T + \theta) \tag{4.88}$$

where $\quad C = N\mu_0\beta^2/k \quad$ and $\quad \theta = \gamma C$

Note that this model indeed gives a susceptibility as required by the experimentally derived equation (4.81).

At low temperatures, the approximation involved in going from (4.84) to (4.85) is no longer justified. In fact, at $T = 0$, the spin system is completely ordered in the sense that all spins on $A$-sites are oriented in parallel and all spins on the $B$-sites are oriented in parallel. Thus, at low temperatures, $\mathbf{M}_a$ amd $\mathbf{M}_b$ are very large, though oppositely directed. As the temperature is raised from zero, the magnetizations $M_a$ and $M_b$ of the two sub-lattices in the absence of a magnetic field vary with temperature in a fashion similar to the spontaneous magnetization of a ferromagnetic material. The spontaneous magnetizations of the sub-lattices disappears at the Néel temperature $T_N$. We shall not discuss here the solution of equation (4.84) for the temperature region below $T_N$; it may suffice to say that the behavior of the experimentally observed susceptibility curve (see Fig. 4.25) can be understood in terms of the model under discussion. However, we may discuss here the occurrence of a Néel temperature for the model. This can be done on the basis of the simpler formulas (4.85), because at the Néel temperature itself the spontaneous magnetization of the sub-lattices vanishes, as can be seen from Fig. 4.23; i.e., at the Néel temperature itself formulas (4.85) should still be valid. Let us investigate then, by putting $H = 0$ in (4.85), for what temperature spontaneous magnetization of the sub-lattices becomes possible. Rewriting (4.85) for

$H = 0$ we obtain

$$M_a + \frac{C}{T}\gamma M_b = 0$$
$$\frac{C}{T}\gamma M_a + M_b = 0 \tag{4.89}$$

For $T > T_N$, these equations have the trivial solutions $M_a = M_b = 0$; i.e., there is no spontaneous magnetization of the sub-lattices above the Néel temperature. If spontaneous magnetization of the sub-lattices is supposed to set in for $T = T_N$, we must require that (4.89) has non-trivial solutions for $M_a$ and $M_b$ at the temperature $T = T_N$. This permits us to calculate $T_N$, because the requirement just stated is equivalent to the requirement that the determinant of the coefficients of $M_a$ and $M_b$ vanishes. Hence

$$\left(\frac{C}{T_N}\gamma\right)^2 = 1 \qquad \text{or} \qquad T_N = C\gamma = \theta \tag{4.90}$$

where the last relationship follows from (4.88). Note that for the model employed here, the Néel temperature turns out to be the same as the paramagnetic Curie temperature $\theta$. (The reader is reminded of the fact that in the ferromagnetic case the simple model employed also gave $\theta_f = \theta$.) According to measurements of $\theta$ and $T_N$, there is a considerable difference between $T_N$ and $\theta$, as may be seen from Table 4.7. This indicates that the model used here was actually too simple. In fact, one can show that if one takes into account antiferromagnetic interactions not only between nearest neighbors, but also between next nearest neighbors, the model above would give $T_N < \theta$. The model discussed here must therefore be considered as an approximation; it predicts the general features of antiferromagnetism correctly, but not the details. It should also be realized that usually the particular body-centered structure assumed in Fig. 4.27 does not apply to the material under study. In Table 4.7 we give values of $T_N$ and $\theta$ for some antiferromagnetic materials.

The question can be raised as to what independent experimental evidence there exists to support the assumption that in an antiferromagnetic material neighboring spins have opposite directions. The answer is that such evidence has been obtained from *neutron diffraction studies*. The neutrons, because of their magnetic moment, can "see" the difference between an "up" spin and a "down" spin and the diffraction patterns (similar to X-ray diffraction patterns) show that the antiparallel spin alignment actually occurs in these materials.

**Table 4.7.** Néel temperature ($T_N$) and paramagnetic Curie temperature ($\theta$) for some antiferromagnetic materials

| Material | $T_N$(°K) | $\theta$(°K) |
|---|---|---|
| $MnF_2$ | 72 | 113 |
| $MnO_2$ | 84 | 316 |
| MnO | 122 | 610 |
| MnS | 165 | 528 |
| FeO | 198 | 570 |
| $NiF_2$ | 73 | 116 |
| CoO | 292 | 280 |

## 4.14 Ferrimagnetic materials

Of the ferrimagnetic materials, the so-called *ferrites* are of greatest interest from the electrical engineering point of view; they behave as ferromagnetic materials in as much as they show spontaneous magnetization below a certain temperature. As far as their conductivity is concerned, they behave as semiconductors. The d-c resistivity of ferrites is many orders of ten higher than that of iron; consequently, the eddy current problem preventing penetration of magnetic flux into the material is much less severe in ferrites than in iron. Ferrites can therefore be used for frequencies up to microwaves in transformer cores and are of great technical importance in this respect.

The chemical formula of simple ferrites may be written as $Me^{2+}Fe_2^{3+}O_4^{2-}$, where $Me^{2+}$ may represent a variety of divalent metallic ions, such as $Fe^{2+}$, $Co^{2+}$, $Mn^{2+}$, $Zn^{2+}$, $Cd^{2+}$, $Mg^{2+}$, etc. Symbolically, one may write the formula as a "mixture" of MeO and $Fe_2O_3$, although a ferrite is actually a solid solution of two such oxides.

Since the oxides contain ions, the magnetic properties should be predictable to a good degree of approximation from the magnetic properties of the ions. Thus, from Table 4.3 we expect, for example, each $Fe^{2+}$ ion to correspond to 4 Bohr magnetons, and each $Fe^{3+}$ to 5 Bohr magnetons. Now, a material such as $Fe^{2+}Fe_2^{3+}O_4^{2-}$ exhibits a saturation magnetization which amounts to $4\beta$ ($\beta = 1$ Bohr magneton) per "molecule" $Fe^{2+}Fe_2^{3+}O_4^{2-}$. It is evident that if the spins of all the ions were lined up in parallel one should find $4 + (2 \times 5) = 14$ Bohr magnetons per molecule. This discrepancy was explained in 1948 by Néel in terms of a model consisting of two sub-lattices, somewhat similar to the *AB* lattice in Fig. 4.27, for which he assumed an antiferromagnetic interaction between *A*-sites and *B*-sites. An important role in this interpretation is played by studies of the atomic arrangement in ferrites, from which it has been possible to identify the $A$- and $B$-sites.

Because of the intimate relationship between the magnetic properties of ferrites and the crystal structures of ferrites, a few remarks may be made here concerning this problem. The oxygen ions in a ferrite form a close-packed face-centered cubic structure. In this arrangement it is found that for every four $O^{2-}$ ions, there are two *octahedral* "holes" (surrounded by six $O^{2-}$ ions) and 1 *tetrahedral* "hole" (surrounded by four $O^{2-}$ ions). The metal ions are distributed over these octahedral and tetrahedral sites. The tetrahedral sites may be identified with the $A$-sites, and the octahedral sites with the $B$-sites mentioned earlier. Thus, the octahedral sub-lattice has twice as many sites as the tetrahedral one. This has been represented schematically in Fig. 4.28. Now, in $Fe^{2+}Fe_2^{3+}O_4^{2-}$

$Fe^{2+}$ $Fe^{3+}$ $B$-sites (octahedral) $4\beta$ $5\beta$ $5\beta$ $A$-sites (tetrahedral) $Fe^{3+}$

**Fig. 4.28.** Schematic representation of $Fe^{2+}$ and $Fe^{3+}$ ions in magnetite. There are two $Fe^{3+}$ ions and one $Fe^{2+}$ ion per molecule of $Fe_3O_4$, as indicated. The net moment per molecule is $4\beta$.

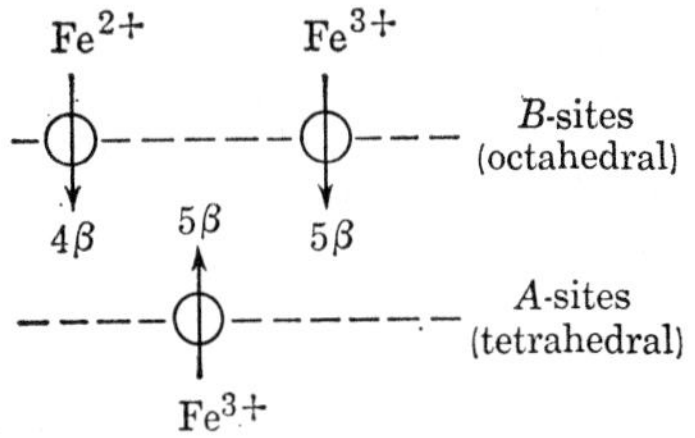

(magnetite), for example, the $Fe^{2+}$ ions occupy half of the octahedral sites; the $Fe^{3+}$ ions occupy the other half of the octahedral sites, and the tetrahedral sites (see Fig. 4.28). Hence, if there exists an antiferromagnetic interaction between $A$- and $B$-sites, we see from Fig. 4.28 that the $Fe^{3+}$ magnetic moments just cancel each other, so that the magnetization of $Fe_3O_4$ should be equal to that produced by the $Fe^{2+}$ ions alone, i.e. $4\beta$ per molecule; that is in agreement with experiment.

The behavior of other ferrites may be explained in similar terms. We may mention here an interesting feature of ferrites, which shows again the importance of the atomic arrangement for the properties of these materials: it is observed that if in $Fe_3O_4$, some of the magnetic $Fe^{2+}$ ions are replaced by non-magnetic ions such as $Zn^{2+}$ or $Cd^{2+}$, the magnetization increases! The reason for this peculiar behavior is the following: Zinc ions go preferably into tetrahedral positions, thereby forcing some of the $Fe^{3+}$ ions from tetrahedral to octahedral sites. Since the $Zn^{2+}$ ions have no magnetic dipole moment, the net magnetization increases, as may be seen from Fig. 4.29. It will be evident that these materials lend themselves,

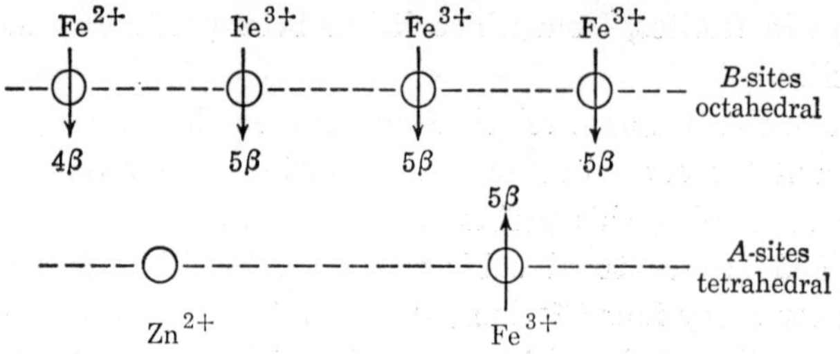

**Fig. 4.29.** Schematic representation of the ionic distribution in magnetite after replacing half of the $Fe^{2+}$ ions by $Zn^{2+}$. The $Zn^{2+}$ ions prefer tetrahedral positions and force $Fe^{3+}$ ions to move into octahedral sites.

within certain limits, to designing materials with prescribed spontaneous magnetization.

## References

L. F. Bates, *Modern Magnetism*, 3d ed., Cambridge, London, 1951.

E. W. Gorter, "Saturation Magnetization and Crystal Chemistry of Ferrimagnetic Oxides," *Philips Research Reports*, **9**, 295, 321, 403 (1954); see also *Proc. IRE*, **43**, 1945 (1955).

J. L. Snoek, *New Developments in Ferromagnetic Materials*, Elsevier, New York, 1947.

J. van den Handel, "Paramagnetism," *Advances in Electronics and Electron Physics*, **6**, 463 (1954).

## Problems

**4.1** A linear conductor carries a current of 10 amperes along the positive $x$-direction. Find the force per meter length on the conductor if it is subjected to a homogeneous flux density of 0.5 weber $m^{-2}$ along the $z$-direction.

**4.2** A linear conductor in air carries a current of 5 amperes; calculate the flux density produced by 1 cm of the conductor in a point at a distance of 1 m normal to the 1 cm section.

**4.3** Show by means of Biot and Savart's law that the flux density produced by an infinitely long straight wire, carrying a current $I$, in a point at a distance $a$ normal to the wire is given by $\mu_0\mu_r I/2\pi a$.

**4.4** Two infinite parallel conductors carry parallel currents of 10 amperes each. Find the magnitude and direction of the force between the conductors per meter length if the distance between them is 20 cm.

**4.5** An electron with velocity vector **v** moves in combined electric and magnetic fields **E** and **B**. Write down the expression for the force on the electron (the "Lorentz force").

**4.6** Show that an electron with a velocity perpendicular to the direction of a homogeneous magnetic field of flux density $B$ describes a circular path with an angular velocity of rotation equal to $eB/m$.

**4.7** A charge of $e$ coulombs is distributed homogeneously over the surface of a sphere of radius $R$ meters. The sphere rotates with an angular velocity $\omega$ about an axis passing through its center. Show that the magnetic dipole moment of the sphere is equal to $\frac{1}{3}e\omega R^2$. Also show that the angular momentum of the sphere is $\frac{2}{3}m\omega R^2$, where $m$ is the total mass of the charge.

**4.8** Consider a charge of $e$ coulombs distributed homogeneously over the surface of a sphere of radius $R$ meters. If the sphere is initially at rest, show that after application of a flux density of $B$ weber m$^{-2}$, the charge distribution will rotate with an angular velocity $\boldsymbol{\omega} = (e/2m)\mathbf{B}$, where $m$ is the total mass of the charge.

**4.9** The magnetic field strength in a piece of copper is $10^6$ ampere m$^{-1}$. Given that the magnetic susceptibility of copper is $-0.5 \times 10^{-5}$, find the flux density and the magnetization in the copper.

**4.10** The magnetic field strength in a piece of $Fe_2O_3$ is $10^6$ ampere m$^{-1}$. Given that the susceptibility of $Fe_2O_3$ at room temperature is $1.4 \times 10^{-3}$, find the flux density and the magnetization in the material; compare the answers with those of the preceding problem. What is the magnetization at the temperature of liquid nitrogen?

**4.11** Consider two point dipoles, each with a strength of 1 Bohr magneton; the dipoles are parallel to each other and parallel to the line joining their centers. If the distance between the dipoles is 2 angstrom, calculate the energy of one dipole in the field of the other and show that the result is equivalent to $kT$ with $T \approx 1°$K. (This shows that ferromagnetic interactions cannot be explained classically, because the interaction energy should be of the order of $kT$ where $T = \theta_f \cong 1000°$K).

**4.12** The saturation value of the magnetization of iron is $1.75 \times 10^6$ ampere m$^{-1}$. Given that iron has a body-centered cubic structure with an elementary cube edge of 2.86 angstroms, calculate the average number of Bohr magnetons contributed to the magnetization per atom.

**4.13** A paramagnetic system of spins is subjected to a homogeneous field of $10^6$ ampere m$^{-1}$ at a temperature of 300°K. Find the average magnetic moment along the field direction per spin in Bohr magnetons. Answer the same question for liquid helium temperature.

# 5

# The Conductivity of Metals

The most characteristic properties of metals are their high electrical and thermal conductivities. For example, the electrical conductivity at room temperature of silver is $0.6 \times 10^8$ ohm$^{-1}$ m$^{-1}$ as compared to $\approx 10^{-16}$ for a good insulator and $2 \times 10^{-2}$ for a semiconductor such as germanium. There is a great deal of evidence to support the notion that the high conductivity in metals is associated with the presence of "free" or "conduction" electrons. These free electrons are able to move throughout the lattice and hence do not belong to particular atoms. The only electrons which have this high degree of freedom are those corresponding to the valence electrons in the atoms. Thus, one may think of a metal as consisting of a lattice of positive ion cores held together by means of a gas of electrons. The electron gas consists of the freely moving valence electrons; the ion cores constitute what is left of the atoms after they have parted with their valence electrons. Although the properties of the electron gas should be discussed on the basis of wave mechanics, the discussion in this chapter will be essentially classical. To this extent, the discussion must be considered as inaccurate. On the other hand, the main purpose here is to arrive at a reasonable qualitative picture concerning the mechanism of conductivity, and this purpose may be achieved by taking a less sophisticated point of view.

The first question we shall consider is this: in terms of the picture just outlined, how should Ohm's law be understood? Then we shall consider problems such as the following: why does the resistivity of metals and alloys increase with increasing temperature; why is the resistivity of an impure metal higher than that of a pure metal; how can we understand Joule's law, which states that the heat developed in a wire is proportional to the square of the voltage across it?

## 5.1 Ohm's law and the relaxation time of electrons

There is probably no law more fundamental to the electrical engineer than Ohm's law. It is used every day in electrical engineering and consequently it seems proper to investigate how this law can be interpreted in terms of an atomic picture. In order to arrive at an atomic interpretation of Ohm's law, we shall formulate the law in the form

$$\mathbf{J} = \sigma \mathbf{E} \tag{5.1}$$

which states that the current density $\mathbf{J}$ is proportional to the field strength $\mathbf{E}$; the proportionality constant $\sigma$ is the *electrical conductivity* of the metal. We shall also refer frequently to the *resistivity*

$$\rho = 1/\sigma \tag{5.2}$$

In connection with the macroscopic description of Ohm's law (5.1) the same remarks hold as were made in connection with the relationship $\mathbf{D} = \epsilon_0\epsilon_r\mathbf{E}$ in section 2.1. Thus, (5.1) holds only if the conductivity is independent of the direction in which it is measured. For polycrystalline materials it is therefore valid, but in the case of single crystals $\sigma$ should in general be replaced by a tensor quantity. For simplicity we shall assume the material to be isotropic. Another remark that should be made here is that when an electric field $\mathbf{E}$ is applied at the instant $t = 0$, the current density requires a non-vanishing period to build up to the equilibrium value $\mathbf{J}$. However, the build-up times are very short (of the order of $10^{-14}$ seconds) and need not concern us at the moment.

One important feature of Ohm's law is the fact that the *current density remains constant in time* as long as $E$ remains constant. Although this feature may seem trivial, it is not so when we consider the problem of conduction from the atomic point of view; in fact, it gives us a clue about the mechanism of conduction as we shall see presently.

Concerning the electron gas, we shall assume that there are $n$ free electrons per $m^3$, where $n$ is presumably of the same order of magnitude as the number of atoms per $m^3$ in the metal. In the absence of an external field, the velocity vectors associated with the electrons at a given instant add up to zero; hence, there is no resultant transport of charge. When an electric field $E$ is applied along the $x$-direction, the electrons of charge

$-e$ are subjected to an acceleration given by

$$a_x = -(e/m)E \tag{5.3}$$

where $m$ is the electronic mass. When $v_x$ represents the velocity component of an electron along the $x$-direction, we may thus write for all electrons

$$\left(\frac{\partial}{\partial t} v_x\right)_{\text{field}} = -\frac{e}{m}E \tag{5.4}$$

The partials and the subscript "field" indicate that this represents the rate of change of $v_x$ due only to the interaction of the electrons with the field; the purpose of introducing this notation will become clear later. Before drawing a conclusion from this argument, let us consider the relationship between the current density $J$ along the $x$-direction and the velocity components $v_x$ of the electrons. Considering a large group of electrons at a certain instant $t$, we may define an average velocity of the electrons in the group as

$$\langle v_x \rangle_t = \frac{1}{N} \sum_{i=1}^{N} v_{xi}(t) \tag{5.5}$$

where the subscript $i$ refers to a particular electron in the group and where $N$ is the number of electrons considered. If there are $n$ electrons per $m^3$, the current density at the instant $t$ may be written as

$$J(t) = -ne\langle v_x \rangle_t \tag{5.6}$$

This relationship follows immediately from Fig. 5.1, in which we have shown a block of the material with a cross section of 1 $m^2$ perpendicular to the $x$-axis. The total charge per second passing through a cross section

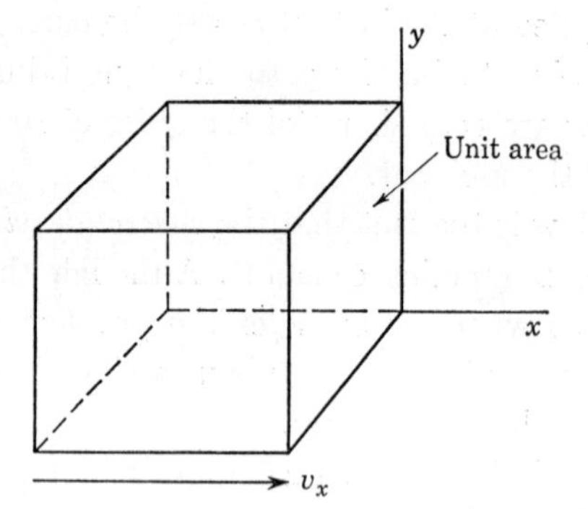


**Fig. 5.1.** Illustrating the statement that the number of particles crossing a unit area per second is equal to the number of particles contained in a volume equal to $v_x$. If the density of particles is $n$ and their charge is $-e$, this gives a current density $J = -nev_x$.

is equal to the charge contained in a block of 1 $m^2$ in cross section and $\langle v_x \rangle$ meters long. (The average velocities in the $y$- and $z$-directions are zero, so for the present argument these may be taken as zero for each of the electrons.) Now, if (5.4) represented the *total* rate of change of $v_x$

for each of the electrons, it would also represent the total rate of change of $\langle v_x \rangle$. But then $\langle v_x \rangle_t$ in expression (5.6) would change at a constant rate, and hence $J(t)$ would change at a constant rate with time. Experiment tells us, however, that, apart from the transient during approximately $10^{-14}$ seconds, $J$ is independent of time for constant $E$. What then is wrong with the atomic interpretation of Ohm's law given so far? It is evident from (5.6) that if $J$ should be constant for a given $E$, the average velocity $\langle v_x \rangle$ must be constant for a given field. This can only mean that expression (5.4) does not give the total rate of change of $\langle v_x \rangle$, and that *there must be other processes which make the total rate of change $d\langle v_x \rangle/dt$ equal to zero.* In other words, what experiment requires is the existence of the following relationship

$$\left[\frac{\partial}{\partial t}\langle v_x \rangle\right]_{\text{field}} + \left[\frac{\partial}{\partial t}\langle v_x \rangle\right]_{\text{other}} = \frac{d}{dt}\langle v_x \rangle = 0 \tag{5.7}$$

which expresses the fact that under influence of an external field we observe a steady value for $J$. The subscript "other" refers to other processes which we have yet to account for. What clues do we have to find these other processes? First of all, if the electrons were in vacuum, there would be no other effect on the electrons but the one produced by the field. One clue then is provided by the fact that the *electrons move in a crystal lattice,* and that the presence of the lattice must account for the "other" processes. Another clue, which points in the same direction, may be derived from the fact that when a current passes through a material, the material is known to warm up. Now, the temperature of a material is a measure of the intensity with which the atoms in the lattice vibrate. Since only the free electrons absorb energy from the field, the extra energy of vibration must have been transferred to the lattice by the electrons. This is illustrated schematically in the block diagram of Fig. 5.2. The

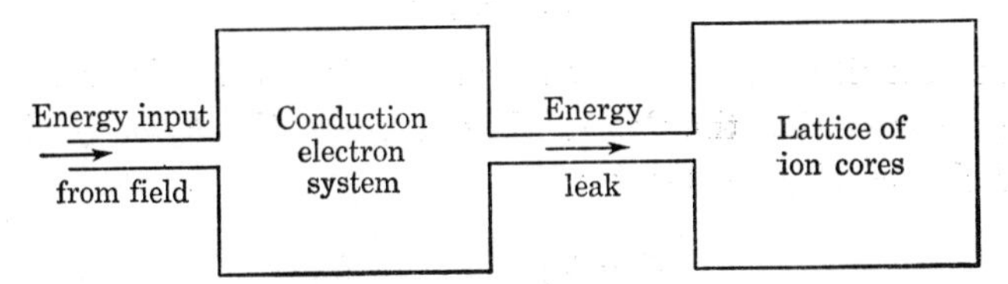


**Fig. 5.2.** Schematic diagram illustrating the system of conduction electrons absorbing energy from the external field, and transferring energy to the lattice of ion cores, where it appears as Joule heat.

small tube connecting the "electron gas" and the "lattice of ion cores" represents the leak through which an excess of energy of the electron gas can be transferred to the lattice. From these arguments we arrive at the conclusion that the "free" electrons in a conductor must interact with the lattice of ion cores, and to this extent they are not completely free. We also conclude that the "other" processes referred to in equation (5.7) must probably be sought in the *electron-lattice interaction*. (It may be mentioned here that collisions between electrons would not solve the dilemma, because such collisions conserve momentum and hence would not change $\langle v_x \rangle$.)

How shall we describe the electron-lattice interaction mathematically? For the moment we shall follow a phenomenological approach; i.e., we shall make an assumption which we consider reasonable, without inquiring about the actual physical mechanism. This assumption is the following: suppose the electrons in a metal have an average velocity $\langle v_x \rangle_0$ at the instant $t = 0$; given that for $t > 0$ there is no external field, the electrons

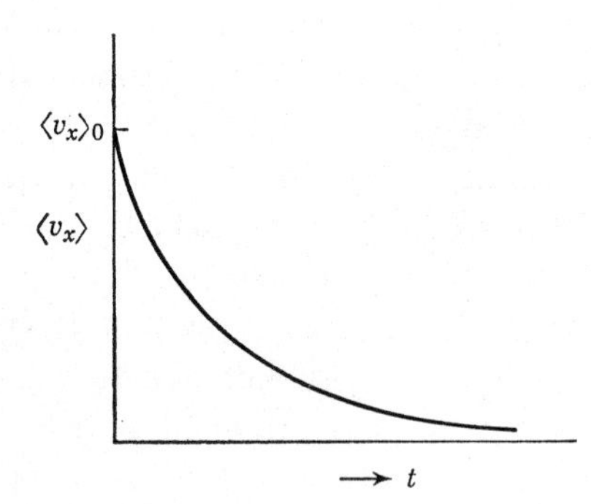


**Fig. 5.3.** Illustrating the influence of collisions alone on the average velocity of a group of particles with an average velocity $\langle v_x \rangle_0$ at the instant $t = 0$. The average velocity is assumed to decay exponentially with time, in accordance with equation (5.8).

interact only with the lattice and ultimately the average velocity will go to zero. We shall assume that under these circumstances the value of $\langle v_x \rangle$ at any instant $t(> 0)$ is given by (see Fig. 5.3).

$$\langle v_x \rangle_t = \langle v_x \rangle_0 e^{-t/\tau} \tag{5.8}$$

where the quantity $\tau$ is called the *relaxation time* of the electrons. In other words, we assume that the electron-lattice interaction *by itself* will produce a rate of change of $\langle v_x \rangle$ given by

$$\left[\frac{\partial}{\partial t}\langle v_x \rangle\right]_{\text{coll}} = -\frac{1}{\tau}\langle v_x \rangle \tag{5.9}$$

The subscript "coll" refers to collisions carried out by the electrons with the lattice. These collision processes will be identified with the "other" processes appearing in equation (5.7).

On the basis of this picture, the conductivity can readily be found in terms of the atomic quantities. Substituting (5.4) and (5.9) into (5.7) we obtain

$$-(e/m)E - (1/\tau)\langle v_x\rangle = 0 \qquad \text{or} \qquad \langle v_x\rangle = -(e\tau/m)E \tag{5.10}$$

The steady average velocity of the electrons in an applied field $E$ is called the *drift velocity* of the carriers in the field. It is noted that the drift velocity is proportional to the field; the absolute magnitude of the proportionality factor

$$\mu_e = e\tau/m \tag{5.11}$$

is called the *mobility* of the electrons. The mobility may thus be defined as the magnitude of the average drift velocity per unit field. Substituting (5.10) into expression (5.6) for the current density, we find

$$J = (ne^2\tau/m)E \tag{5.12}$$

and it follows then from (5.1) that the measurable quantity $\sigma$ is related to the atomic quantities by means of

$$1/\rho = \sigma = ne^2\tau/m \tag{5.13}$$

Thus, the experimental demonstration of a steady flow of current in a metal subjected to a constant field has compelled us to introduce an interaction between the electrons and the atomic lattice. This leaves us with the problem of the atomic interpretation of the relaxation time which describes this interaction. The question may also be raised as to whether a relaxation time can be defined at all, but this subject lies outside the scope of this book.

## 5.2 Relaxation time, collision time, and mean free path

In the preceding section we introduced the concept of a relaxation time on phenomenological grounds. In this section we shall see that such a relaxation time may be obtained from a picture in which one assumes the electrons to collide with "obstacles"; the nature of these obstacles will be discussed in subsequent sections.

Suppose an electron moves with a velocity $v_x$ along the $x$-direction of a Cartesian coordinate system. Let us assume that the electron collides with an obstacle in such a way that the probability for it to be scattered over an angle between $\theta$ and $\theta + d\theta$ is given by $P(\theta)2\pi \sin\theta\, d\theta$, where $P(\theta)$ is an arbitrary function of the scattering angle $\theta$ [see Fig. 5.4(a)]. Evidently then, we should have

$$2\pi \int_{\theta=0}^{\pi} P(\theta) \sin\theta\, d\theta = 1 \tag{5.14}$$

We shall assume that the magnitudes of the velocities before and after

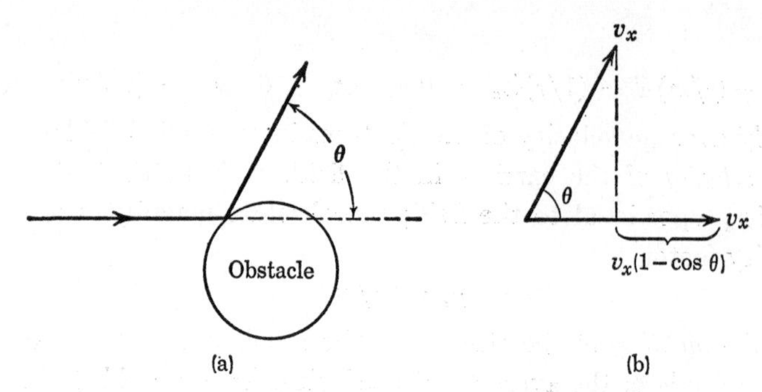


**Fig. 5.4.** Schematic representation of scattering of an electron by an obstacle over an angle $\theta$ is given in (a); (b) shows that as a result of this scattering process, the $x$-component of the velocity has been reduced by the amount $v_x(1 - \cos\theta)$, assuming that the magnitude of the velocity is the same after scattering as before.

scattering are the same. In connection with this assumption we should comment that actually some energy is transferred in the conduction process from the electrons to the lattice (represented in the present picture by the obstacles); this is, after all, the reason for the dissipation of Joule heat, as referred to in Fig. 5.2. However, for all cases of interest to the subject matter discussed in this book one can show that the percentage change in velocity resulting from the collisions is very small indeed. In discussing the collision process, therefore, we may assume equal velocities before and after the collision. Under these circumstances, an electron scattered over an angle $\theta$ has a velocity component along the $x$-direction equal to $v_x \cos\theta$. Hence, it has lost along the $x$-direction a velocity component amounting to (see Fig. 5.4b)

$$v_x(1 - \cos\theta) \tag{5.15}$$

Consider now a large group of electrons, all moving with the same velocity $v_x$ along the $x$-direction. Assuming all these particles carry out one collision with an obstacle, what is the average velocity component lost by the particles? The answer is evidently

$$v_x(1 - \langle\cos\theta\rangle) \tag{5.16}$$

where $\langle\cos\theta\rangle$ is the average value of the cosine of the scattering angle. Since the probability for scattering over a given angle is described by $P(\theta)$, we must have

$$\langle\cos\theta\rangle = 2\pi \int_{\theta=0}^{\pi} P(\theta) \cos\theta \sin\theta \, d\theta \tag{5.17}$$

Note that if the scattering is spherically symmetric, i.e. if $P(\theta)$ is constant, $\langle\cos\theta\rangle = 0$.

Let us now consider the collisions from the point of view of time. Assume that the probability for an electron to collide with an obstacle during a short time interval $dt$ is given by $dt/\tau_c$, where $\tau_c$ is a constant with the dimensions of time. Let $F(t)$ represent the probability that an electron moves for a time $t$ without suffering a collision, and let $F(t + dt)$ represent the same quantity for a period $t + dt$. We may then write

$$F(t + dt) = F(t) + \frac{dF}{dt}\, dt \tag{5.18}$$

on the other hand, we may also write

$$F(t + dt) = F(t)F(dt) = F(t)[1 - dt/\tau_c] \tag{5.19}$$

in view of the definition of $\tau_c$. Hence, the last two equations give

$$dF/dt = -F/\tau_c \qquad \text{or} \qquad F(t) = e^{-t/\tau_c} \tag{5.20}$$

since $F(t) = 1$ for $t = 0$, which fixes the integration constant. What is the average time which elapses between collisions? The answer is

$$\langle t \rangle = -\int_0^\infty t\, \frac{dF}{dt}\, dt \tag{5.21}$$

because $-(dF/dt)\, dt$ gives the probability that the electron collides between $t$ and $t + dt$. Substituting $F(t)$ from (5.20) into (5.21) we find

$$\langle t \rangle = \tau_c \tag{5.22}$$

Hence, the physical meaning of $\tau_c$ introduced above is that it represents the *average time between collisions*.

We are now prepared to answer the question: what is the rate of change of $\langle v_x \rangle$ for a group of electrons resulting from collisions with the obstacles described earlier? In accordance with (5.16) an electron loses per collision on the average

$$\langle v_x \rangle (1 - \langle \cos \theta \rangle)$$

Since the number of collisions carried out per second is on the average equal to $1/\tau_c$, we have

$$\left[\frac{\partial}{\partial t} \langle v_x \rangle\right]_{\text{coll}} = -\frac{\langle v_x \rangle}{\tau_c} (1 - \langle \cos \theta \rangle) \tag{5.23}$$

Comparing (5.23) with expression (5.9) which defines the relaxation time, we see that the relaxation time is related to the mean time between collisions as follows:

$$\tau = \frac{\tau_c}{1 - \langle \cos \theta \rangle} \tag{5.24}$$

When the scattering is isotropic, $\langle \cos \theta \rangle = 0$ and $\tau = \tau_c$. If the scattering takes place predominantly in the direction of incidence, $\langle \cos \theta \rangle$ will have a positive value, and the relaxation time $\tau > \tau_c$. This is understandable qualitatively, because the electrons then will "remember" longer what their initial direction of motion was, and consequently the time required to make the velocity distribution random increases.

The mean free path of an electron, $\lambda$, will be defined as

$$\lambda = v\tau_c \tag{5.25}$$

where $v$ is the total velocity of an electron. It will be evident that if the obstacles are hard spheres, the quantity $\lambda$ would be determined by the concentration of obstacles. Thus, in such a case the collision time $\tau_c$ would be inversely proportional to the velocity $v$ of the electrons for a given concentration of obstacles. Consequently, when $\langle \cos \theta \rangle$ is independent of the velocity of the electrons, the relaxation time will be inversely proportional to $v$. This raises an important question, because in the preceding section we introduced a single relaxation time $\tau$, without any reference to the distribution of electron velocities. In the case of metals, however, it can be shown that the relaxation time $\tau$ which determines the conductivity always refers to a single total velocity, viz., the one corresponding to the *Fermi energy*. In order to explain this properly, one would have to carry through a procedure involving Fermi statistics, a subject which lies outside the scope of the present treatment. A few qualitative remarks may suffice here. The free electrons in a metal can accept a series of discrete energy levels. At absolute zero, all energy levels below a certain value $W_F$ are filled, all those above $W_F$ being empty; $W_F$ is called the Fermi level of the electrons. The magnitude of $W_F$ is of the order of several electron volts. At any temperature $T$, the probability of a state corresponding to the energy $W$ being occupied by an electron is given by

$$f(W) = \frac{1}{\exp[(W - W_F)/kT] + 1} \tag{5.26}$$

For $T = 0$, this function is a step function, as indicated in Fig. 5.5; for $T > 0$, one obtains a curve which differs slightly from the one correspond-

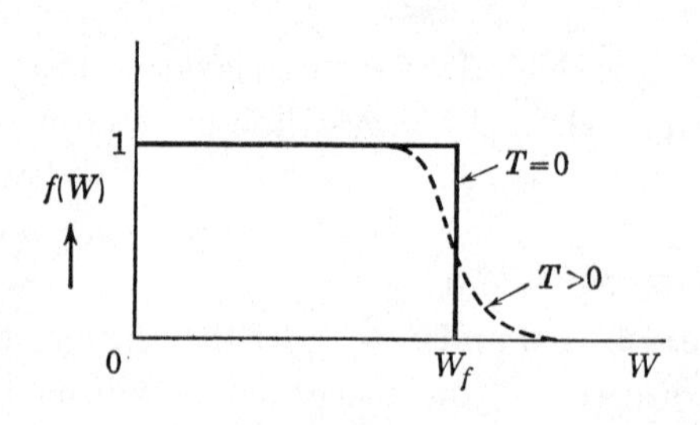


**Fig. 5.5.** Schematic representation of the Fermi distribution function (5.26) for $T = 0$ and for a temperature $T > 0$. Actually, the width of the energy region over which the dashed curve differs appreciably from the curve for $T = 0$ is very small compared to $W_f$ for practical temperatures.

ing to $T = 0$, as indicated in an exaggerated fashion in Fig. 5.5. Since $kT$ at room temperature is only 0.025 ev, $kT$ is very small compared to $W_F$, and, consequently, the energy region over which temperature influences the energy distribution of the electrons is very narrow. The velocity of the electrons with the Fermi energy, $v_F$, is given by

$$\tfrac{1}{2}mv_F^2 = W_F \tag{5.27}$$

For $E_F = 5$ ev, which is a representative figure, one finds $v_F \cong 4 \times 10^6$ meters per second (1 ev $= 1.6 \times 10^{-19}$ joules). The relaxation time appearing in the theory of conductivity of metals refers to electrons with a velocity $v_F$.

We have seen then that it is possible to interpret the relaxation time in terms of collisions of the electrons with obstacles. Our next task will be to determine what these obstacles are, and if the model so obtained agrees with the observed dependence of the conductivity on temperature, impurity concentration, etc.

## 5.3 Electron scattering and the resistivity of metals

Formula (5.13) derived in section 5.1 permits us to calculate $\tau$ from experimental values of the conductivity, or at least to obtain an estimate of order of magnitude. If one assumes for monovalent metals that the number of free electrons per m³ is equal to the number of atoms per m³, one obtains at 0°C values for $\tau$ as given in Table 5.1. It is observed that

**Table 5.1.** CONDUCTIVITY ($\sigma$), FERMI LEVEL ($W_F$), MEAN FREE PATH ($\lambda$), AND RELAXATION TIME ($\tau$) AT 0°C FOR SOME MONOVALENT METALS

| Metal | $\sigma$ ($10^8$ ohm$^{-1}$ m$^{-1}$) | $W_F$ (ev) | $\lambda$ (angstroms) | $\tau$ ($10^{-14}$ sec) |
|---|---|---|---|---|
| Li | 0.12 | 4.7 | 110 | 0.9 |
| Na | 0.23 | 3.1 | 350 | 3.1 |
| K | 0.19 | 2.1 | 370 | 4.4 |
| Cu | 0.64 | 7.0 | 420 | 2.7 |
| Ag | 0.68 | 5.5 | 570 | 4.1 |

the relaxation times are of the order of $10^{-14}$ seconds. Since this number may not convey to the reader any more than the fact that it is small, it is useful to consider the results in terms of the mean free path defined by (5.25). It was mentioned in the preceding section that $\tau$ in metals refers to electrons with the Fermi velocity $v_F$. Thus, if we assume that the scattering in these metals is isotropic (it can be shown that this is a good approximation for the temperatures of interest), we have $\tau = \tau_c$

and $\lambda = v_F\tau$. The values of $\lambda$ so obtained are also indicated in Table 5.1. It is noted that at 0°C, the mean free path for scattering is several hundred angstroms. We may mention further that the resistivity of metals at low temperatures decreases to very small values, as may be seen from the example in Fig. 5.6 for sodium. Although some of the assumptions made

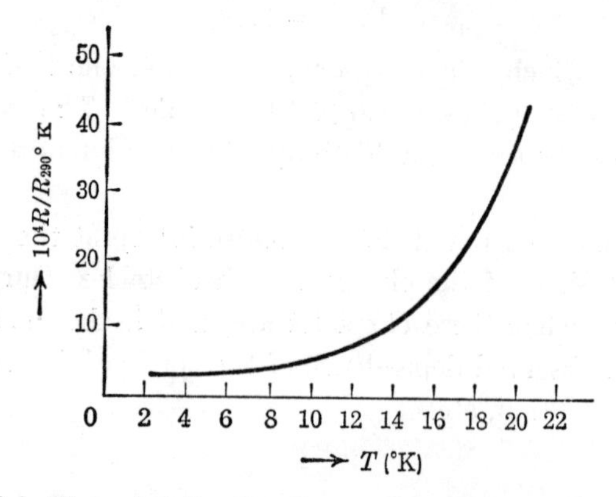


**Fig. 5.6.** The resistivity of sodium as a function of temperature in the low temperature region. The constant value approached at low temperatures is due to lattice imperfections and impurities, and corresponds to $\rho_i$ in the text.

above are not applicable for liquid helium temperatures, we may state that the *resistivity of a perfect, pure single crystal of a metal goes to zero as T approaches zero.* In other words, at extremely low temperatures, $\lambda$ reaches macroscopic values. From these arguments one can draw an important, though somewhat negative conclusion, viz., that the *obstacles* introduced earlier *are not the individual ion cores* in the metallic lattice. In fact, if we were to picture the conduction electrons as moving between the positive ion cores, one would expect, in terms of classical physics, mean free paths of the order of a few angstroms, as indicated in Fig. 5.7. This classical picture thus breaks down, and we have to look for another one. The reason for the breakdown of the classical picture lies in the wave aspect of

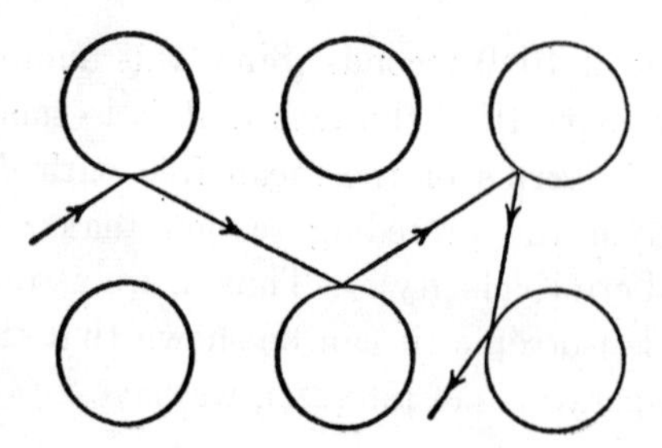

**Fig. 5.7.** Schematic representation of the classical picture of scattering of an electron by the ion cores; this picture predicts a mean free path of a few angstroms, in contradiction with experimental information.

elementary particles. In fact, it can be shown theoretically that when the wave nature of the electrons is taken into account properly, *an electron moving in a perfectly periodic lattice would not be scattered at all.* In other words, the waves representing a moving electron in a perfect crystal resemble the electromagnetic waves corresponding to visible light which passes undeviated through a crystal such as sodium chloride. Thus, the present interpretation of the electrical resistivity of metals is based on the notion that *scattering of the electrons is due to deviations from perfect periodicity of the lattice.* In pure metals, such deviations may be caused, for example, by the vibrations of the atoms in the lattice; they may also be caused by atoms which occupy "wrong" positions in the lattice, or by lattice sites which are unoccupied, or by impurity atoms. Thus, there are various kinds of obstacles which we have to consider in the scattering of electrons in metals, each of them giving rise separately to a certain relaxation time $\tau_j$. Since the probability for scattering by a particular kind of obstacle per unit time is equal to $1/\tau_j$ (for isotropic scattering) the resultant relaxation time may be obtained from the relation

$$\frac{1}{\tau} = \sum_j \frac{1}{\tau_j} \tag{5.28}$$

since scattering probabilities are additive. Consider then a metal with a given concentration of imperfections, i.e. a given concentration of impurities and dislocated atoms. Let us assume that the relaxation time associated with all these imperfections is equal to $\tau_i$. The resultant relaxation time may then be written as

$$1/\tau = 1/\tau_i + 1/\tau_T \tag{5.29}$$

where $\tau_T$ represents the relaxation time resulting from the thermal vibrations of the lattice. When the concentration of imperfections is independent of the temperature $T$, $\tau_i$ will presumably be independent of temperature (because $\lambda_i$ is fixed and $v_F$ is independent of $T$, for all practical purposes). The resistivity of the metal may then be represented in accordance with formula (5.13) by

$$\rho = \frac{m}{ne^2}\left(\frac{1}{\tau_i} + \frac{1}{\tau_T}\right) \tag{5.30}$$

The only temperature-dependent quantity appearing in this formula is $\tau_T$ (the number of electrons, $n$, in a metal is independent of $T$!). This formula then should explain the observed *positive temperature coefficient of the resistivity of metals* (we are ignoring some very special cases). As $T$ increases, the amplitude of the atomic vibrations increases; hence the deviations from periodicity become more severe and the scattering by lattice vibrations increases. Consequently, as $T$ increases, $\lambda$ decreases and since $v_F$ is constant, $\tau_T$ decreases. Thus, as the temperature of a metal (or alloy) is raised, the electrical resistivity increases. At temperatures above the so-called Debye temperature of the material, the energy of the lattice vibrations increases linearly with $T$, and consequently, the resistivity increases linearly with $T$ in this region, obeying the formula

$$\rho = \rho_i + aT \tag{5.30a}$$

where $\rho_i$ and $a$ are constants. The separation of the resistivity into a temperature-independent part and a temperature-dependent part is referred to as Matthiessen's rule. The Debye temperature may be considered as a measure for the maximum frequency of the lattice vibrations; hard materials have higher Debye temperatures than soft materials, as may be seen from the values given in Table 5.2 below. The Debye temperature plays an important role in the specific heat of a material in that above the Debye temperature the specific heat is approximately constant, whereas below it, the specific heat drops rapidly, approaching zero as $T \to 0$.

**Table 5.2.** Debye temperature ($\theta_D$) for some metals, in degrees absolute

| Metal | Ag | Al | Au | Cu | Fe | Pb |
|---|---|---|---|---|---|---|
| $\theta_D$ | 215 | 398 | 180 | 315 | 420 | 88 |

As temperatures fall below the Debye temperature, $\tau_T$ increases rapidly and as $T$ goes to zero, $\tau_T$ approaches infinity. There exists a close *relationship between the resistivity and the specific heat* of the lattice vibrations; in fact, the Debye temperature may be calculated from the resistivity versus temperature curve of a metal. Since the electrical resistivity at very low temperatures resulting from scattering of electrons by lattice vibrations becomes negligible, the total resistivity is essentially determined by the imperfections and impurities. Thus, with reference to Fig. 5.6, the constant resistivity near 4°K is a measure for the content of imperfections and impurities in the sodium sample.

The arguments given also make the following observation intelligible: when one adds to a metal like nickel a small amount of copper, the conductivity decreases, notwithstanding the fact that copper has a higher conductivity than nickel. The reason is, of course, that in replacing some of the nickel atoms in the lattice, the copper atoms destroy the lattice periodicity in their immediate neighborhood and thus act as new scattering centers for the conduction electrons.

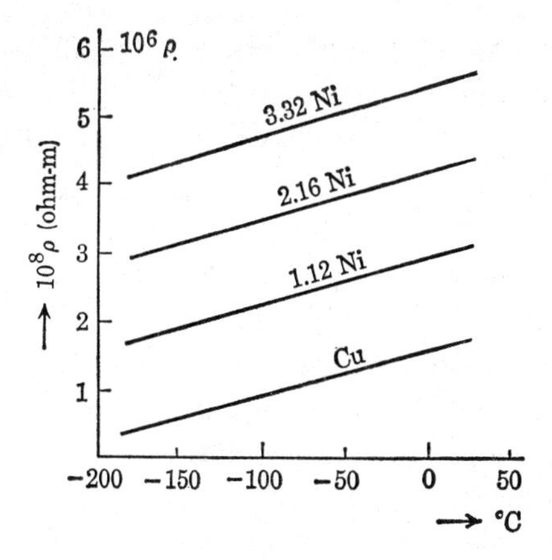


**Fig. 5.8.** The resistivity of alloys of nickel in copper as a function of temperature; the numbers indicate the atomic percentage of nickel in the alloy.

In Fig. 5.8 we give an example of the influence of temperature and of impurity content on the resistivity of copper. It is noted that $\rho$ varies linearly with $T$ in accordance with (5.30a), and that the resistivity of the alloys increases with increasing nickel content.

## 5.4 The heat developed in a current-carrying conductor

In this section we shall give an elementary derivation of the well-known experimental fact that the heat developed per $m^3$ per second in a conductor carrying a current density $J$ as a result of an applied field $E$ is given by

$$W = JE = \sigma E^2 \tag{5.31}$$

If $J$ is in amperes per $m^2$ and $E$ is in volts per m, $W$ is found in watts per $m^3$. Consider a particular electron which at the instant $t = 0$ has carried out a collision with the lattice; at that instant let the velocity components of the electron be $v_x$, $v_y$, $v_z$. Assuming that at the instant $t(> 0)$ the electron has not yet collided again, and assuming the field $E$ is applied along the *negative* $x$-direction, the velocity components of the electron at the time $t$ will be given by

$$v_x + (e/m)Et, \qquad v_y, \qquad \text{and} \qquad v_z \tag{5.32}$$

The increment $(e/m)Et$ results from the acceleration along the $x$-direction produced by the field on the electron of charge $-e$. The increase in energy of the electron over the period $t$ is therefore

$$(\Delta W)_t = \tfrac{1}{2}m[(2e/m)Ev_xt + (e/m)^2E^2t^2] \tag{5.33}$$

When this expression is averaged over a large number of electrons which have all lived through the period $t$ without having suffered a collision, but which presumably have a random distribution of their velocities, one finds

$$\langle \Delta W \rangle_t = \tfrac{1}{2}m[(e/m)^2 E^2 t^2] = e^2E^2t^2/2m \tag{5.34}$$

because $\langle v_x \rangle$ is zero for such a group.

Now, according to (5.20), the probability that an electron will live through the period $t$ without making a collision is equal to $F(t) = \exp[-t/\tau_c]$. For simplicity we shall assume that the electrons are scattered isotropically, so that the average time between collisions, $\tau_c$, is equal to the relaxation time $\tau$. Also, the probability that an electron will suffer a collision during a time interval $dt$ is given by $dt/\tau_c = dt/\tau$. Hence the product $e^{-t/\tau}(dt/\tau)$ gives us the probability that the electron makes a collision between $t$ and $t + dt$. Consequently, the average energy increase of the electrons during the period between two collisions is equal to

$$\langle \Delta W \rangle = \int_{t=0}^{\infty} \langle \Delta W \rangle_t e^{-t/\tau} \frac{dt}{\tau} \tag{5.35}$$

where $\langle \Delta W \rangle_t$ is given by (5.34). Substituting (5.34) into (5.35) and carrying out the integration, one finds

$$\langle \Delta W \rangle = \tau^2 e^2 E^2/m \tag{5.36}$$

Assuming there are $n$ electrons per m$^3$, and assuming that these electrons transfer their excess energy to the lattice, one obtains for the total energy dissipated per m$^3$ per second

$$W = \frac{n}{\tau} \langle \Delta W \rangle = \left( \frac{ne^2\tau}{m} \right) E^2 = \sigma E^2 \tag{5.37}$$

in agreement with the experimental relation (5.31).

## 5.5 The thermal conductivity of metals

When a homogeneous isotropic material is subjected to a temperature gradient, a flow of heat results in the direction opposite to the gradient. Thus, when $dT/dx$ represents the temperature gradient and $Q$ represents the heat flow density, one defines the thermal conductivity $K$ by means of the equation

$$Q = -K(dT/dx) \tag{5.38}$$

When $Q$ is expressed in watts per m$^2$, and $dT/dx$ in degrees C per meter, one finds $K$ in watt m$^{-1}$ degree$^{-1}$. It is well known that, at least at normal temperatures, metals conduct heat much better than insulators. In Table 5.3 we have given values for $K$ at room temperature for various materials to

**Table 5.3.** THERMAL CONDUCTIVITY $K$ AT ROOM TEMPERATURE IN WATT METERS$^{-1}$ DEGREES$^{-1}$, FOR VARIOUS MATERIALS

| Material | $K$ | Material | $K$ |
|---|---|---|---|
| Al | 209 | Mica | |
| | | ($\perp$ cleavage plane) | 0.75 |
| Constantan | | Quartz | |
| (60 Cu, 40 Ni) | 22.5 | ($\parallel$ axis) | 12.6 |
| Cu | 385 | ($\perp$ axis) | 6.7 |
| Fe | 67 | Carbon | 4.2 |
| Ni | 54 | Glass (window) | 1.0 |

illustrate the range of thermal conductivities encountered. In insulating solids, the heat current is carried by the lattice vibrations. This, in part, is also the case in metals, but the thermal conductivity due to the conduction electrons predominates in these materials. We shall consider only the *thermal conductivity arising from the conduction electrons.*

Assuming a temperature gradient $dT/dx$ along the $x$-direction, the average energy of the electrons will also be a function of $x$. Thus, let $W(0)$ be the average energy per electron in the plane $x = 0$. At a distance $\Delta x$ from this plane, the energy of the electrons is then given by

$$W(\Delta x) = W(0) + \frac{dW}{dx}\Delta x = W(0) + \frac{dW}{dT}\frac{dT}{dx}\Delta x \tag{5.39}$$

The atomic picture of heat conduction by electrons is the following: One considers a unit area of the plane at $x = 0$, and calculates the net transport of energy through that plane resulting from electrons coming from the two regions $x > 0$ and $x < 0$. It is assumed that the energy of an electron which passes through the plane at $x = 0$ is equal to the average electron energy at $\Delta x$, where $\Delta x$ gives the location along the $x$-direction where the electron last collided with the lattice. Thus, let $v_{xi}$ represent the velocity component of a particular electron along the $x$-direction, and let $-\Delta x_i$ represent the location of the plane where it last collided. If there were $n_i$ such electrons per m$^3$, the number passing the plane $x = 0$ per second would be $n_i v_{xi}$. Since the energy carried by electron $i$ is obtained by replacing $\Delta x$ by $-\Delta x_i$ in (5.39), the total heat current density through the plane $x = 0$ in the positive $x$-direction may be written as

$$Q = \sum_i \left[ W(0) - \frac{dW}{dT}\frac{dT}{dx}\Delta x_i \right] v_{xi} \tag{5.40}$$

where the summation extends over all electrons per m$^3$. Since the average velocity of the electrons per unit volume is zero, the sum containing the quantity $W(0)$ vanishes. We are thus left with

$$Q = -\sum_i \frac{dW}{dT}\frac{dT}{dx} v_{xi}\,\Delta x_i = -n\frac{dW}{dT}\frac{dT}{dx}\langle v_x\,\Delta x\rangle \tag{5.41}$$

where $n$ stands for the total number of conduction electrons per m³, and where $\langle v_x\,\Delta x\rangle$ is the average value of $v_x\,\Delta x$. Now, an electron moving in a direction making an angle $\theta$ with the $x$-axis has a velocity component $v_x = v\cos\theta$, where $v$ is the total velocity. Similarly, if $\Delta x$ measures the distance from the plane $x = 0$ to the plane where it collided last, the elec-

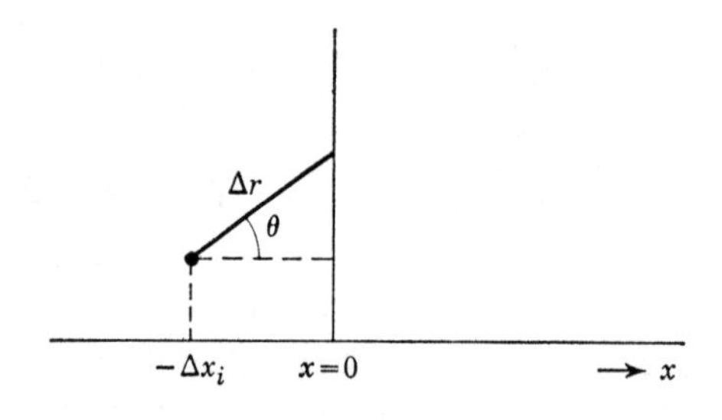


**Fig. 5.9.** Illustrating the geometry used in the derivation of the heat conductivity.

tron has actually covered a path length $\Delta r = \Delta x/\cos\theta$ between the instant it collided and the instant it arrives at the plane $x = 0$, as indicated in Fig. 5.9. We may thus write (5.41) in the form

$$Q = -n\frac{dW}{dT}\frac{dT}{dx}\langle v\,\Delta r\cos^2\theta\rangle = -\frac{1}{3}n\frac{dW}{dT}\frac{dT}{dx}\langle v\,\Delta r\rangle \tag{5.42}$$

where the last equality follows from the fact that

$$\langle\cos^2\theta\rangle = \frac{\int_0^\pi \cos^2\theta\sin\theta\,d\theta}{\int_0^\pi \sin\theta\,d\theta} = \frac{1}{3} \tag{5.43}$$

At this point we shall make a statement that will not be proved here, but which bears relation to an earlier remark in connection with the electrical conductivity of metals: from a more rigorous treatment it can be shown that only the mean free path or the relaxation time corresponding to electrons with the Fermi energy is of importance. Thus $v$ in (5.42) may be replaced by the velocity $v_F$ (Fermi velocity), and $\Delta r$ by $\lambda$; here, $\lambda$ is the mean free path for scattering corresponding to electrons with the velocity $v_F$. Hence,

$$Q = -\frac{1}{3}n\frac{dW}{dT}\frac{dT}{dx}v_F\lambda \tag{5.44}$$

The quantity $n(dW/dT)$ represents the energy increase of the electron gas per m³ per degree; i.e., this quantity is simply the *specific heat of the electron gas at constant volume, C.* Since $\lambda = v_F\tau$, where $\tau$ is the relaxation time, we have

$$Q = -\frac{1}{3} C v_F^2 \tau \frac{dT}{dx} \tag{5.45}$$

From Fermi-Dirac statistics it can be shown that the specific heat of the electron gas is equal to

$$C = \frac{n\pi^2 k^2 T}{m v_F^2} \tag{5.46}$$

The last two equations thus give

$$Q = -\frac{1}{3} \frac{n\pi^2 k^2 T\tau}{m} \frac{dT}{dx} \tag{5.47}$$

Comparing this expression with (5.38) we find for the thermal conductivity

$$K = \frac{1}{3} \frac{n\pi^2 k^2 T\tau}{m} \tag{5.48}$$

We can draw several conclusions from this result. First, we may estimate $K$ from (5.48) and compare the result with the values given in Table 5.2. For $T = 300°\text{K}$, $n \cong 5 \times 10^{28}\ \text{m}^{-3}$, $\tau \cong 10^{-14}$ seconds, $k = 1.38 \times 10^{-23}$ joule degree$^{-1}$, $m = 9 \times 10^{-31}$ kg, we find $K \cong 100$ watt m$^{-1}$ degree$^{-1}$, which is of the correct order of magnitude. How does the thermal conductivity depend on temperature? All quantities appearing in expression (5.48) for $K$ are independent of temperature, except $T$ and $\tau$. As mentioned in Section 5.3, $\tau$ varies as $T^{-1}$ for temperatures above the Debye temperature. Hence, in this temperature region the *thermal conductivity is independent of temperature.* At low temperatures the behavior is more complicated, as illustrated in Fig. 5.10 for copper.

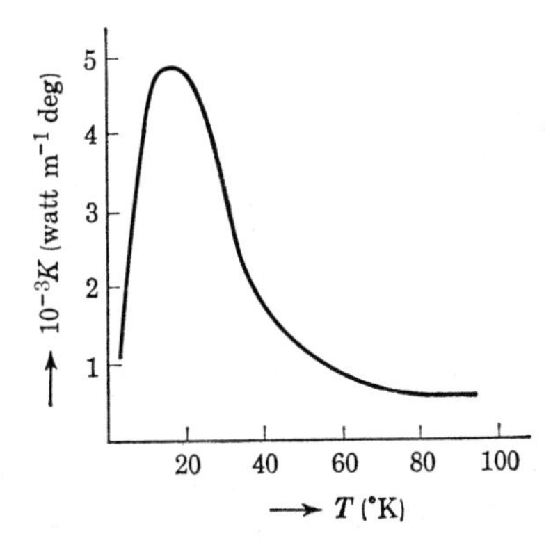


**Fig. 5.10.** The thermal conductivity of copper at low temperatures. [After R. Berman and D. K. C. MacDonald, *Proc. Roy. Soc.* (London) **A211**, 122 (1952)]

There exists an interesting relationship between the electrical conductivity and the thermal conductivity of a metal. According to expressions (5.13) and (5.48) we may write

$$\frac{K}{\sigma} = \frac{\pi^2}{3}\frac{k^2}{e^2}T \quad \text{or} \quad \frac{K}{\sigma T} = \frac{\pi^2}{3}\frac{k^2}{e^2} \equiv L \tag{5.49}$$

The theory thus predicts that $K/\sigma T$ for all metals should be equal to a universal constant; this law is known as the *Wiedemann-Franz law.* The constant represented by $L = 2.45 \times 10^{-8}$ watt ohm degree² is known as the *Lorenz number.* In Table 5.4 we give experimental values of the Lorenz number for various metals, at 0°C and at 100°C. It is observed that the agreement between theory and experiment is satisfactory. It should be remarked here that the contribution to $K$ from the lattice vibrations has been neglected in our considerations. At low temperatures, the theory given does not apply, and a relaxation time for the thermal conductivity and electrical conductivity *cannot* be defined uniquely. Thus, the Lorenz number for copper near 15°K is an order of magnitude smaller than at room temperature. The Wiedemann-Franz law holds only above the Debye temperature.

**Table 5.4.** Experimental Lorenz number for various metals at 0°C and at 100°C, expressed in $10^{-8}$ watt ohm degree²

| Metal | 0°C | 100°C | Metal | 0°C | 100°C |
|---|---|---|---|---|---|
| Ag | 2.31 | 2.37 | Pb | 2.47 | 2.56 |
| Au | 2.35 | 2.40 | Pt | 2.51 | 2.60 |
| Cu | 2.23 | 2.33 | Mo | 2.61 | 2.79 |

## 5.6 Superconductivity

A phenomenon which has defied a satisfactory atomic interpretation for many years is that of superconductivity. On the other hand, phenomenological theories based on thermodynamics (Gorter and Casimir) and on electrodynamics (London) have provided a certain amount of insight into the relationships between the various physical properties of superconductors. Also, new attempts to find an atomic interpretation made by Fröhlich and by Bardeen since 1950 appear to be promising, and it is not impossible that ultimately superconductivity can be understood within the framework of the modern theory of metals. A discussion of the various aspects of this subject falls outside the scope of this book and the interested reader is referred to the references provided at the end of this chapter. On the other hand, since superconductors have recently found applications in switching elements called *cryotrons*, it seems proper to provide at least a few experimental facts about superconductivity here.

Superconductivity was discovered in 1911 by Kamerlingh Onnes in Leiden when he observed that the electrical resistivity of mercury disappeared completely at temperatures below approximately 4.2 degrees Kelvin. As shown in Fig. 5.11, the resistance of the specimen measured

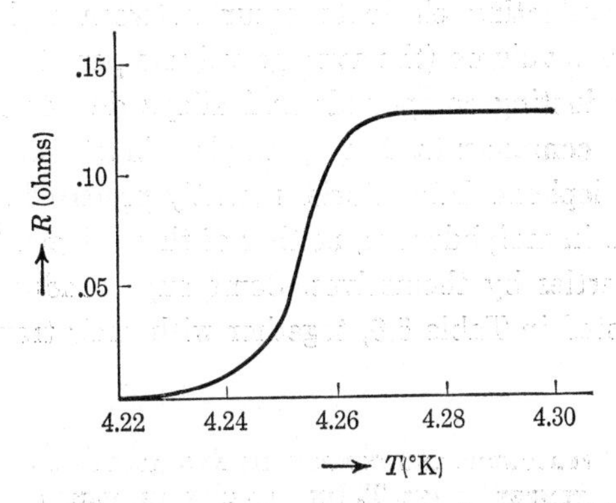


**Fig. 5.11.** The resistance of mercury as a function of temperature as observed by Kamerlingh Onnes in 1911.

by Onnes drops to zero within a very narrow temperature interval of only a few hundredths of a degree. Consequently, it is meaningful to introduce a *transition temperature*, $T_c$, at which the transition from the "normal" state to the superconducting state occurs. Since the discovery of the superconducting properties of mercury, many elements and some thirty compounds and alloys have been found which exhibit the same behavior. The superconducting elements and their transition temperatures are listed in Table 5.5. It is noted that the transition temperatures all lie

**Table 5.5.** SUPERCONDUCTING ELEMENTS AND THEIR TRANSITION TEMPERATURES, $T_c$ (IN DEGREES ABSOLUTE)

| Element | $T_c$ | Element | $T_c$ | Element | $T_c$ |
|---|---|---|---|---|---|
| Al | 1.14 | Ru | 0.47 | Re | 1.0 |
| Ti | 0.53 | Cd | 0.54 | Os | 0.71 |
| V | 5.1 | In | 3.37 | Hg | 4.12 |
| Zn | 0.79 | Sn (white) | 3.69 | Tl | 2.38 |
| Ga | 1.07 | La | 4.71 | Pb | 7.26 |
| Zr | 0.7 | Hf | 0.35 | Th | 1.32 |
| Nb | 9.22 | Ta | 4.38 | U | 0.8 |

below 10°K. It is, of course, possible that other elements become superconducting at sufficiently low temperatures, since measurements have not usually been carried out below about 0.07°K. It is interesting that the elements which at room temperature are good conductors (Cu, Ag,

Au and the alkali metals) are absent from the list of superconducting elements. In fact, the superconducting elements are relatively poor conductors at room temperature. One notices also the absence from Table 5.5 of the ferromagnetic elements Fe, Ni, Co. It may be mentioned here that the superconducting elements occur between rather well defined limits of the atomic volume (the average volume per atom).

The superconducting compounds and alloys do not necessarily have superconducting components. For example, Matthias and coworkers (1956) at Bell Telephone Laboratories recently prepared superconducting alloys of rhodium in molybdenum, neither of these elements having superconducting properties by themselves. Some superconducting compounds and alloys are listed in Table 5.6, together with their transition temperatures.

**Table 5.6.** Some superconducting compounds and alloys and their transition temperatures, $T_c$ (in degrees absolute)

| Compound | $T_c$ | Compound | $T_c$ |
|---|---|---|---|
| $Pb_2Au$ | 7.0 | NbN | 14.7 |
| $PbTl_2$ | 3.8 | MoN | 12.0 |
| SnSb | 3.9 | NbB | 6 |
| CuS | 1.6 | ZrC | 2.3 |

The question may be raised as to whether the resistivity of a superconductor is actually zero or just very small. In reply to this question one may state that in an unpublished experiment by Professor Collins at M.I.T., an induced current of several hundred amperes has been observed to flow in a superconducting lead ring for over a year, without a detectable change. It thus seems that $\rho = 0$ may be considered as one of the characteristic properties of the superconducting state.

The transition temperature of a superconductor can be reduced by

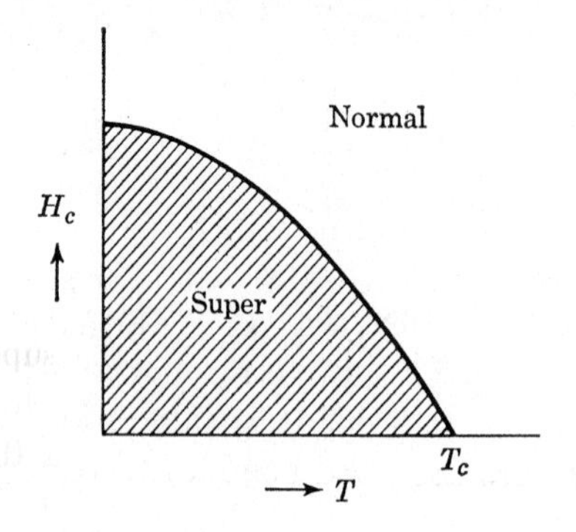


**Fig. 5.12.** Schematic representation of the critical magnetic field $H_c$ as a function of temperature.

the application of a magnetic field. With reference to Fig. 5.12, suppose a superconductor has a temperature $T < T_c$. If a magnetic field $H$ is applied, the material remains superconducting until a critical field $H_c$ is reached such that for $H > H_c$, the material is in the normal state. The curve in Fig. 5.12 represents schematically the functional relationship between the critical field $H_c$ and the temperature of the superconductor. The transition from the superconducting to the normal state under influence of a magnetic field is reversible. The function $H_c(T)$ follows with good accuracy a formula of the form

$$H_c(T) = H_0\left(1 - \frac{T^2}{T_c^2}\right) \tag{5.50}$$

where $H_0$ and $T_c$ are constants characteristic of the material. Experimental $H_c(T)$ curves for a number of elements are given in Fig. 5.13.

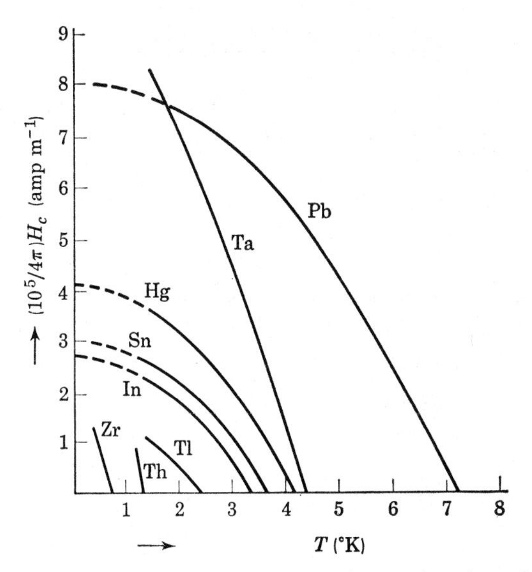


**Fig. 5.13.** The critical magnetic field $H_c$ as a function of temperature for various elements.

The magnetic field which causes a superconductor to become normal is not necessarily an externally applied field; it may also arise as a result of electric current flow in the conductor. Thus, superconductivity in a long circular wire of radius $r$ may be destroyed when the current $I$ exceeds the value $I_c$, which at the surface of the wire would produce the critical magnetic field $H_c$. Hence, according to expression (4.8), the critical current $I_c = 2\pi r H_c$; this is referred to as *Silsbee's rule.* This rule prevents the use of superconductors as coils for the production of strong magnetic fields.

The disappearance of superconductivity for fields above the critical field is the principle on which a cryotron operates. With reference to Fig. 5.14, consider a wire made of a superconducting material $A$, inside a coil of a superconducting material $B$. Let the temperature of the system

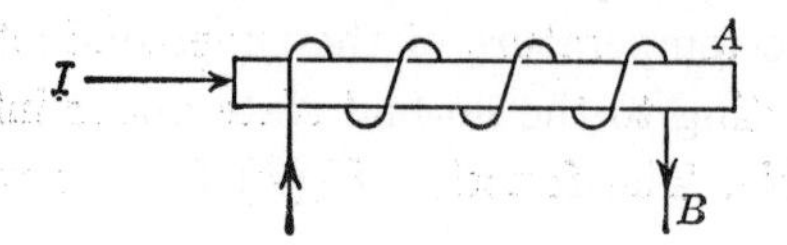


**Fig. 5.14.** A single cryotron in which the current $I$ in $A$ is controlled by the current in $B$.

be below the transition temperatures of the two materials, so that both are superconducting. The current in the central wire $A$ can then be controlled by a current in the coil because the magnetic field produced by the latter may exceed the critical field of the core material at the operating temperature. The control current required to make the core "normal" depends on the d-c current flowing through the core because this current also produces a magnetic field. A metal such as tantalum is useful as a core material if the operating temperature is that of a liquid helium bath (4.2°K). The reason is, that for tantalum, $T_c = 4.38$ so that the critical field required to produce the normal state is relatively small (see Fig. 5.13). The coil material must be chosen so that it remains superconducting even if the control current flows; niobium or lead are thus suitable coil materials. The single cryotron represented in Fig. 5.14 may be used as an element in a more complicated device such as a flip-flop. Since the elements are superconductors, the power consumption is very low and it has been estimated that a large-scale digital computer could be built requiring no more than $\frac{1}{2}$ watt (not counting the helium cryostat and terminal equipment!). Also, since the elements can be made small, the essential parts of such a computer would occupy only one cubic foot.

So far, we have discussed only one aspect of superconductivity, viz., the fact that $\rho = 0$. An important experiment was carried out in 1933 by Meissner and Ochsenfeld which showed that the magnetic flux density in a superconductor also vanishes, i.e., $\mathbf{B} = 0$. This result does not follow from the fact that $\rho = 0$, as can be seen in the following way: if $\rho = 0$, the electric field $\mathbf{E}$ in a superconductor carrying a non-vanishing current must also vanish. In accordance with one of Maxwell's equations we have curl $\mathbf{E} = -d\mathbf{B}/dt$ so that, for $\mathbf{E} = 0$, this predicts $d\mathbf{B}/dt = 0$, but *not* $\mathbf{B} = 0$. In other words, the Meissner-Ochsenfeld experiment requires the characterization of a superconductor by *two* relations: $\rho = 0$ and $\mathbf{B} = 0$. For a detailed exposition of the consequences of these two relations for the macroscopic behavior of superconductors the reader is referred to London's book cited below. Note that according to formula (4.24) the equation $\mathbf{B} = 0$ means that the magnetization in a superconductor $\mathbf{M} = -\mathbf{H}$. Consequently, since in general $M = (\mu_r - 1)H$ (see expression 4.23), we may also state that *for a superconductor the relative permeability is zero;* this is referred to as *perfect diamagnetism.*

With reference to the atomic interpretation of superconductivity, the results of experiments by Maxwell, and by Reynolds and coworkers, (1950) may be mentioned here. Making use of isotopes of mercury, they observed a decrease in the transition temperature $T_c$ with increasing isotopic mass $M$ such that

$$M^{1/2}T_c = \text{constant} \tag{5.51}$$

The same relationship has since been found for tin and lead isotopes. It is important to realize that a change in the isotopic mass leaves the electronic structure of the material unaltered; i.e., the relationship (5.51) indicates that the phenomenon of superconductivity is determined in some way by the interaction between the electrons and atoms in the superconductor. In the preceding sections we have seen that the normal resistivity of a metal is determined in part by the interaction between electrons and lattice vibrations. Since the vibrational frequency of an elastically bound particle is proportional to the square root of the mass of the particle, expression (5.51) suggests strongly that superconductivity may also result from the interaction of electrons and lattice vibrations. This idea was actually advanced before the experiments mentioned were carried out, and forms the basis of the theories of Frölich and Bardeen.

## References

J. Bardeen, "Electrical Conductivity of Metals," *J. Appl. Phys.* **11**, 88 (1940).

D. A. Buck, "Cryotron—a superconductive circuit element," *Proc. I.R.E.* **44**, 482 (1956).

F. London, *Superfluids*, Wiley, New York, 1950, vol. 1.

N. F. Mott and H. Jones, *Theory of the Properties of Metals and Alloys*, Oxford, New York, 1936.

D. Shoenberg, *Superconductivity*, Cambridge University Press, 2nd ed., 1952.

## Problems

**5.1** A copper wire is 1 m long and has a uniform cross section of 0.1 mm$^2$; the resistance of the wire at room temperature is found to be 0.172 ohm. What is the resistivity of the material?

**5.2** A uniform silver wire has a resistivity of $1.54 \times 10^{-8}$ ohm m at room temperature. For an electric field along the wire of 1 volt cm$^{-1}$ compute the average drift velocity of the electrons, assuming there are $5.8 \times 10^{28}$ conduction electrons per m$^3$. Also calculate the mobility and relaxation time of the electrons.

**5.3** Assuming the Fermi energy for silver is 5.5 ev, find the velocity of an electron with the Fermi energy. Compare the answer with that for the drift velocity in the preceding problem. What is the mean free path for scattering in the silver wire of the preceding problem?

**5.4** A copper wire has a resistivity of $1.8 \times 10^{-8}$ ohm m at room temperature (300°K). Assuming the copper is very pure, estimate the resistivity at 700°C and the percentage change in the resistivity from room temperature to 700°C.

**5.5** Nichrome is a well-known alloy of nickel, iron, and chromium used for electric furnace windings. It has a resistivity of $1.0 \times 10^{-6}$ ohm m at room temperature, and heating to a temperature of 700°C increases its resistivity by approximately 7 percent. Compare this information with that obtained in problem 5.4 and explain the difference in behavior of nichrome and copper. Assuming Matthiessen's rule holds for nichrome, find the resistivity due to impurity scattering alone.

**5.6** If one plots the resistivity of disordered alloys of gold in copper as a function of the atomic concentration of gold, one obtains a dome-shaped curve with the following features: it starts at $1.7 \times 10^{-8}$ ohm m for pure copper, then rises to a maximum value of $14 \times 10^{-8}$ ohm m, occurring for equal atomic concentrations of the two elements, and then decreases to the value of $2.0 \times 10^{-8}$ ohm m for pure gold. Explain the shape of this curve qualitatively.

**5.7** Find the probability for an electronic state to be occupied at room temperature if the energy of this state lies 0.1 ev above the Fermi level. Do the same for a state which lies 0.1 ev below the Fermi level.

**5.8** In the Fermi distribution function introduce a new variable $x = (W - W_F)/kT$; $x$ measures the energy in units of $kT$ relative to the Fermi energy $W_F$. Show that $-dF/dx$ is a symmetrical function with a maximum at $x = 0$. On the basis of the information so obtained, show that $\partial F/\partial W$ has an appreciable value only in an energy region of the order of $kT$ on either side of the Fermi level.

**5.9** A specimen of pure annealed copper has a resistivity of $1.56 \times 10^{-8}$ ohm m at 300°K. If nickel is added to copper, the resistivity increases by $1.25 \times 10^{-8}$ ohm m per added atomic percent nickel. Similarly, silver increases the resistivity of copper by $0.14 \times 10^{-8}$ ohm m per added atomic percent. For an alloy of 0.2 atomic percent nickel and 0.4 atomic percent silver in copper, what is the theoretical resistivity of the alloy at 300°K and at 4°K? Also calculate the percentage increase in the resistivity per degree centigrade for the alloy and for pure copper at 300°K.

**5.10** Given that the relaxation time of the conduction electrons in copper is equal to $2.7 \times 10^{-14}$ seconds, compute the average increase in the $x$-component of the velocity between two collisions when an electric field of 1 volt $\text{cm}^{-1}$ is applied in the negative $x$-direction. Compare this increase with the actual velocity of an electron with the Fermi energy ($W_F = 7.0$ ev for copper). What is the average increase in energy of the electrons between two collisions?

**5.11** The temperature difference between the inside and outside of a glass window is 72° Fahrenheit. The glass has a thermal conductivity of 0.0025 calories $\text{sec}^{-1}$ $\text{cm}^{-1}$ $\text{degree}^{-1}$ and is 1 mm thick. Find the energy loss in joules through the window per $\text{m}^2$ per hour.

**5.12** The electrical resistivities of copper and kanthal (an alloy of iron, chromium and aluminum) at room temperature are respectively $1.7 \times 10^{-8}$ and $1.4 \times 10^{-6}$ ohm m. Assuming the Wiedemann-Franz law holds for these materials, find the *electronic* contributions to the thermal conductivities of these materials. (In alloys such as kanthal, the contribution to $K$ from the lattice vibrations is comparable to that from the electrons.)

**5.13** For pure metals, the theory predicts a thermal conductivity which is practically independent of temperature above the Debye temperature. Why is this so? In contrast, consider an alloy such as nichrome or kanthal in which the scattering of electrons due to "impurities" predominates over scattering by lattice vibrations. What temperature dependence do you predict for the electronic thermal conductivity in these alloys?

**5.14** Suppose the conduction electrons in a metal are subjected to an alternating electric field $E = E_0 \cos \omega t$ applied along the $x$-direction. Let $v_x$ represent the average $x$-component of the velocity of the electrons and let $\tau$ be their relaxation time. Set up the equation of motion for an electron and show that

$$v_x(t) = -\frac{e\tau E_0}{m}\left[\frac{\cos \omega t}{1 + \omega^2\tau^2} + \frac{\omega\tau \sin \omega t}{1 + \omega^2\tau^2}\right]$$

where $m$ is the electronic mass. Note that the first term in brackets is in phase with the field, whereas the second lags 90 degrees behind the field.

From this result, show that the frequency dependence of the conductivity is given by $\sigma(\omega) = \sigma_{\text{static}}/(1 + \omega^2\tau^2)$.

**5.15** Consider the conduction electrons in a metal subjected to an alternating electric field produced by light waves. Since $\tau \approx 10^{-14}$ seconds, $\omega\tau \gg 1$ for such fields, and the equation of motion for an electron is simply $m(d^2x/dt^2) = -eE_0 \cos \omega t$, where $x$ represents the displacement resulting from a field $E_0 \cos \omega t$ along the $x$-direction. What is the polarization $P$ in the metal if there are $N$ free electrons per unit volume? From this, show that the relative dielectric constant is given by

$$\epsilon_r = 1 - (Ne^2/\epsilon_0 m\omega^2)$$

(If $\epsilon_r > 0$, the metal is transparent to light at normal incidence; if $\epsilon_r < 0$ one can show that total reflection occurs. When applied to sodium, one finds that it is transparent in the ultraviolet for wavelengths $\lambda \lesssim 2100$ angstroms, in agreement with experiment.)

# 6

# The Mechanism of Conduction in Semiconductors

## 6.1 Classifying materials as semiconductors

In principle, the interpretation of the electrical conductivity of semiconductors is similar to that for metals. However, as a result of the difference in the nature of the chemical bond in these materials as compared to the metallic bond, the details are more complicated. The first question which arises is: which materials are classified as semiconductors, and on what basis are they classified as such? In order to answer this question, let us first consider Table 6.1 in which the electrical resistivity of a number of materials is given at 20°C.

**Table 6.1.** Electrical resistivity of various materials at 20°C, in ohm meter

| Metals | | Semiconductors | | Insulators | |
|---|---|---|---|---|---|
| Ag.......... | $1.6 \times 10^{-8}$ | Ge ("pure") | 0.47 | glass | $10^{10}-10^{11}$ |
| Cu.......... | $1.7 \times 10^{-8}$ | Si ("pure") | 3000 | mica | $9 \times 10^{14}$ |
| Al.......... | $2.8 \times 10^{-8}$ | $Fe_3O_4$ | $\sim 0.01$ | diamond | $10^{14}$ |
| Fe.......... | $10 \times 10^{-8}$ | InSb | $\sim 2 \times 10^{4}$ | | |
| Constantan.. | $49 \times 10^{-8}$ | Sn (gray) | $\sim 2 \times 10^{-6}$ | | |
| Nichrome.... | $100 \times 10^{-8}$ | | | | |

It is observed that at this temperature, the materials classified as semiconductors have a resitivity somewhere between that of a typical metal and that of a typical insulator. However, since the resistivity of semiconductors in general depends strongly on temperature, this kind of classification is not altogether satisfactory. In fact, at very low temperatures, say near liquid helium temperatures, a semiconductor may become indistinguishable from an insulator as far as the magnitude of the resistivity is concerned. A better criterion for characterizing semiconductors may be established by considering Fig. 6.1 which gives a schematic plot of resistivity versus temperature for two typical cases encountered in practice. It is observed that these materials exhibit a *negative temperature coefficient of the resistivity,* at least over part of the temperature range. In the preceding chapter we have seen that metals, on the other hand, have a positive $d\rho/dT$ for all temperatures (except in some very special cases). The occurrence of a negative $d\rho/dT$ is therefore a generally accepted basis for classifying a material as a semiconductor. At this point it should be remarked that insulators also have a negative $d\rho/dT$, and that in this respect they behave as semiconductors. The difference between semiconductors and insulators is therefore only of a *quantitative* nature, as will be seen more clearly later; insulators and semiconductors differ only in the magnitudes of their resistivity.

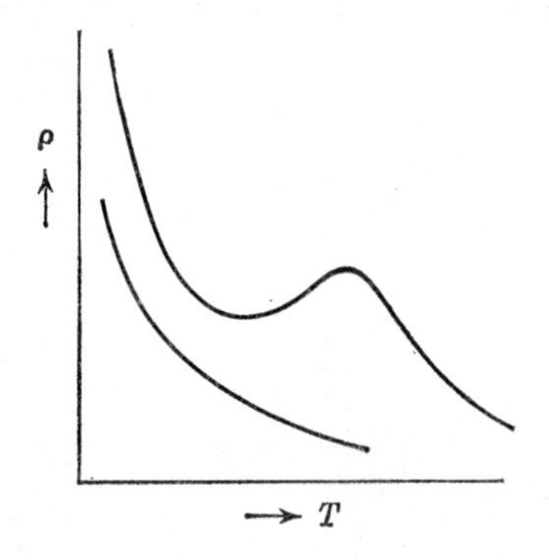


**Fig. 6.1.** Two typical forms of the resistivity of a semiconductor as function of temperature.

Although silicon and germanium are the most popular semiconductors, there are, of course, many others. In Table 6.2 we have indicated the semiconducting elements as they appear in the periodic system; these materials are called *elemental semiconductors.* There are also a large number of compound semiconductors, such as metallic-oxides and sulfides. Many of these are of great practical importance. For example, PbS is used in photoconductive devices, BaO in oxide coated cathodes, cesium antimonide in photomultipliers, etc. Within the scope of the present volume it is impossible to discuss the properties of all these materials, and we shall be interested in general principles rather than in specific materials. Because of the importance of germanium and silicon in present-day electrical engi-

**Table 6.2.** The A subgroups of the 3rd, 4th, 5th, 6th, and 7th columns of the periodic system of elements; the elements enclosed by the line are semiconductors

| IIIA | IVA | VA | VIA | VIIA |
|---|---|---|---|---|
| B | C | N | O | F |
| Al | Si | P | S | Cl |
| Ga | Ge | As | Se | Br |
| In | Sn | Sb | Te | I |
| Tl | Pb | Bi | Po | At |

neering, we shall use these materials to illustrate the behavior of elemental semiconductors; some general remarks on compound semiconductors will be made as the discussion warrants them.

## 6.2 The chemical bond in Si and Ge and its consequences

Silicon and germanium are elements of the fourth column in the periodic table. In the solid state they crystallize into what is known as the *diamond*

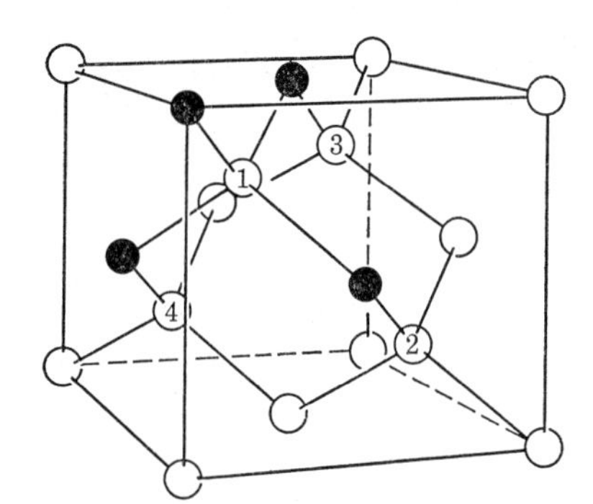


**Fig. 6.2.** The crystal structure of diamond, showing each atom surrounded by four others at the corners of a regular tetrahedron. The lattice consists of a face-centered cubic arrangement of atoms plus four atoms inside the cube; the latter are numbered 1, 2, 3, 4. The four atoms which are bound to atom No. 1 are shown in black.

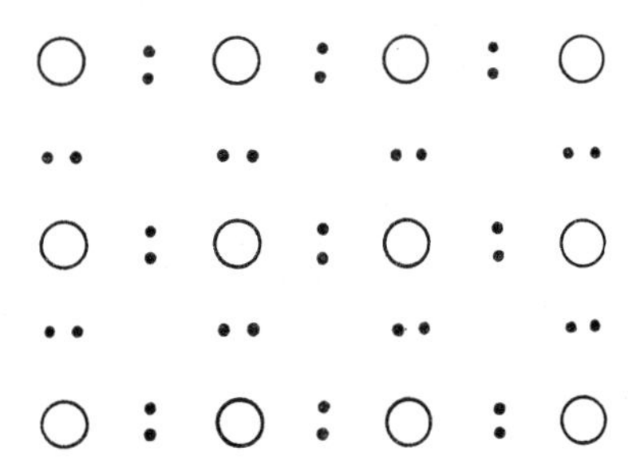

**Fig. 6.3.** Schematic two-dimensional representation of the electron-pair bonds in the diamond structure; the dots represent electrons. In the actual three-dimensional structure, the electron-pair bonds emerge from a given atom to the corners of a regular tetrahedron, the angles between the bonds being approximately 109°.

*structure*, represented in Fig. 6.2. In this structure, a given atom is surrounded by four others occupying the corners of a regular tetrahedron. The bonds between a given atom in this structure and its neighbors are called *electron-pair bonds* because bonds of this type are accomplished by pairs of electrons, each partner contributing one electron to each bond. Thus, since each atom has four valence electrons, it has just enough to provide for electron-pair bonds with four other atoms. A two-dimensional schematic picture of the electron-pair bonds is given in Fig. 6.3. In the actual three-dimensional lattice, the bonds make angles of approximately 109 degrees with each other, in a fashion similar to the four bonds in a molecule such as $CH_4$. Bonds of this kind, in which the atoms share electrons with each other are referred to as *homopolar*, in contrast with *heteropolar* or ionic bonds.

In a pure silicon or germanium crystal at absolute zero, all valence electrons take part in the homopolar bonds mentioned; i.e., *there are no electrons which have the freedom to move through the crystal.* This situation may be contrasted with that encountered in metals, in which there is a constant number of "free" electrons at all temperatures. Thus, at absolute zero, silicon and germanium are insulators, simply because there are no mobile charge carriers available. Suppose now that the temperature is raised; i.e., the vibrations carried out by the atoms in the lattice are made more violent. Some of the valence electrons may then absorb a sufficient amount of energy from the lattice vibrations to be released from the bonds, and once set free they can move through the crystal and contribute to the conductivity in an applied electric field. It is worthwhile at this point to investigate the consequences of the presence of a broken bond. With reference to Fig. 6.4 suppose an electron is taken away from a particular

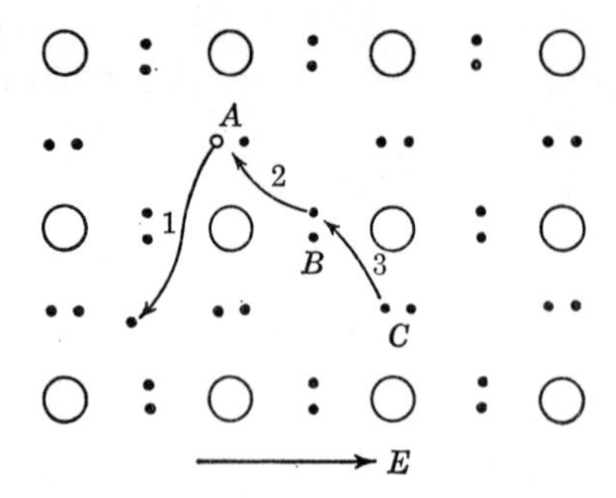


**Fig. 6.4.** Schematic illustration of the consequences of breaking an electron-pair bond $A$; the open circle at $A$ represents a hole. By the successive electron jumps 1, 2, 3, the hole is created and moves in the direction of the field; the released electron drifts in a direction opposite to the field.

bond $A$, as indicated. We are then left with a bond which contains only one electron instead of two. Suppose now that the material is subjected to an electric field as indicated. An electron from a bond such as $B$ may then, under influence of the field, tend to move into the vacant electronic state in $A$; this leaves a vacant electronic state in bond $B$ and an electron from another bond such as $C$ may fill this vacant state. By repeating this process we see that charge is transported under influence of an electric field by virtue of the *absence* of an electron in an electron-pair bond. Note that *the vacant electronic state moves in the same direction as would a positive charge carrier.* For this reason, these vacant electronic states are called *positive holes,* or simply *holes.* We have thus come to the qualitative conclusion that as the temperature of the material is raised from zero to some temperature $T > 0$, a number of the bonds will be broken and conduction may be observed as a result of the motion of electrons and holes under influence of an external field.

For the pure elements Ge and Si under discussion, it is obvious from what has been said above that the number of "free" electrons must equal the number of mobile holes. Semiconductors of this kind are called *intrinsic semiconductors.* They must be distinguished from *extrinsic semiconductors,* in which the charge carriers (electrons or holes) are present as a result of *impurities* built into the crystal.

When a "free" electron moves about in the crystal it may encounter one of the broken bonds mentioned earlier. It is then possible that the electron recombines with the hole, and consequently two charge carriers are lost for the conduction process. On the other hand, any of the normal bonds may generate a free electron and a hole by absorption of energy from the lattice vibrations. Thus, at a given temperature $T$ one will ultimately establish an equilibrium situation in which the number of electron-hole pairs generated by thermal motion per second equals the number of electron-hole pairs lost per second due to recombination. At a given temperature then, there will be a certain average number of electrons and an equal average number of holes in the material. As the temperature is raised, the average number of free charge carriers increases. This process of generation and recombination may be written in the form of a chemical reaction as follows:

$$\text{bound electron} \leftrightarrows \text{free electron} + \text{hole} \tag{6.1}$$

The extent to which this reaction proceeds to the right at a given temperature depends on the chemical bond. If the electrons are bound strongly, the number of charge carriers will be relatively small.

## 6.3 The density of carriers in intrinsic semiconductors; the energy gap

In this section we shall put the qualitative picture outlined in the preceding section on a more quantitative basis. This is usually done in terms of a so-called *energy band picture*, which arises from the following arguments. We have seen that at absolute zero, all valence electrons in a material such as germanium take part in the bonds between the atoms. We have also seen that such a valence electron may become free upon absorption of a sufficient amount of energy. This implies that the energy of a free electron is higher than that of a bound valence electron, and schematically we may thus consider the simple energy diagram represented in Fig. 6.5.

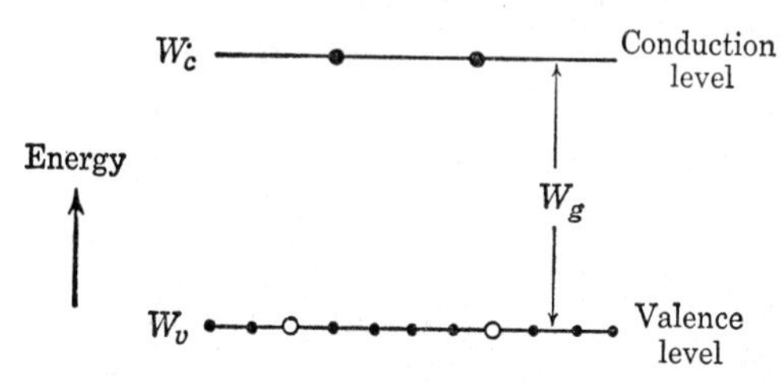


**Fig. 6.5.** Schematic illustration of the energy of an electron taking part in the valence bonds (dots in the valence level $W_v$) and the energy of electrons moving freely through the crystal ($W_c$). At absolute zero, all states in the $W_v$ level are occupied by electrons and all those in the $W_c$ level are empty. At temperatures different from zero, a fraction of the valence electrons is thermally excited into the conduction level (dots in $W_c$ level), leaving holes in the valence level indicated by the open circles.

In this diagram it has been assumed that all valence electrons have the same energy $W_v$, whereas all free electrons have been assumed to have an energy $W_c$; the subscript "*c*" refers to "conduction" electrons, which is the usual name for the free electrons in semiconductors. We shall refer to the $W_c$ level as the "conduction level," and to $W_v$ as the "valence level." The black dots in Fig. 6.5 represent electrons, the open circles represent holes. At absolute zero, let there be $N$ electrons per m³ in the valence level and, of course, no electrons in the conduction level. Given the energy difference

$$W_g = W_c - W_v \tag{6.2}$$

how does the number of conduction electrons vary with temperature? In equilibrium at a temperature $T$, let there be $n$ conduction electrons per m³ and an equal density of holes in the intrinsic semiconductor under discussion. In the preceding section we have seen that conduction electrons may recombine with holes. The probability that such a recombination process will occur is proportional to the number of electron-hole collisions; thus, the number of recombination processes per m³ per second will be proportional to the density of electrons ($n$) and to the density of holes ($n$). Hence, we may write

$$\left(\frac{\partial n}{\partial t}\right)_{\text{recombination}} = -rn^2 \tag{6.3}$$

where $r$ is a constant; the minus sign indicates a decrease in $n$ resulting from recombination. The rate at which electron-hole pairs are created by absorption of thermal energy is presumably proportional to the density of electrons which are available for excitation, $(N - n)$, and to a Boltzmann factor exp $(-W_g/kT)$, where $W_g$ is the energy gap between the valence and conduction levels. Thus,

$$\left(\frac{\partial n}{\partial t}\right)_{\text{thermal excitation}} = G(N - n)e^{-W_g/kT} \tag{6.4}$$

where $G$ is a constant. Since in equilibrium the sum of (6.3) and (6.4) must vanish ($\delta n/\delta t = 0$), we conclude that

$$rn^2 = G(N - n)e^{-W_g/kT} \tag{6.5}$$

As long as $n \ll N$, which is usually the case, one finds for the density of conduction electrons

$$n = \text{constant} \times e^{-W_g/2kT} \tag{6.6}$$

Note that the density of conduction electrons (which in our case equals the hole density) is determined by the ratio of *half* the gap width and $kT$; the factor of $\frac{1}{2}$ enters as a result of the $n^2$ in the recombination rate.

The picture given here is inaccurate in the sense that in a solid, the valence electrons do not all have the same energy; nor do the conduction electrons all have the same energy. In fact, if atoms are brought together to form a solid, the discrete atomic energy levels broaden into a band of very closely spaced energy levels; the broadening is a result of the interaction between the atoms. The same is true for the energy levels which are available to the conduction electrons. A more accurate electron energy level picture for intrinsic semiconductors and insulators is represented in Fig. 6.6. On the basis of certain simplifying assumptions concerning the relation between the energy and wavelength of the electrons in the valence band and in the conduction band, one again finds for $n$ an expression of the same form as (6.6). The constant can be evaluated from statistical mechanics under these circumstances, and as shown in monographs on

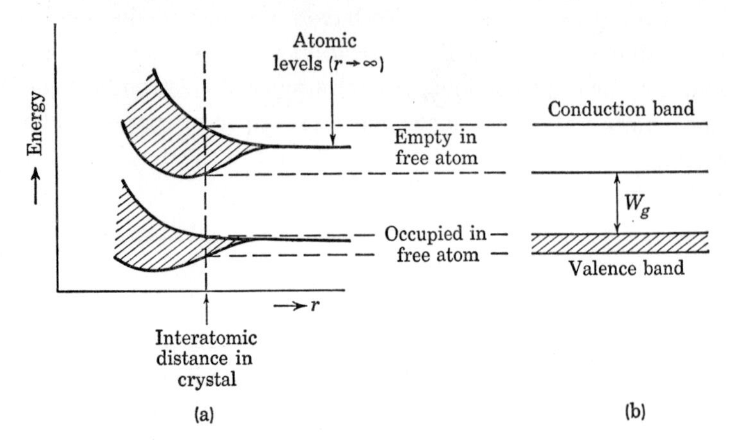


**Fig. 6.6.** In (a) we have represented schematically how the electronic energy levels broaden into bands when the atoms are brought close together. In the crystalline state, the energy band picture presented in (b) is derived from (a) as indicated, leading to a band of energies for the valence electrons (valence band), and a band of energies corresponding to electrons moving about in the crystal (conduction band).

semiconductors one finds

$$n = \left(\frac{2\pi mkT}{h^2}\right)^{3/2} \left(\frac{m_e m_h}{m^2}\right)^{3/4} T^{3/2} e^{-W_g/2kT} \tag{6.7}$$

Here, $h$ is Planck's constant ($= 6.62 \times 10^{-34}$ joule sec); $m$ is the mass of an electron in free space. The quantities $m_e$ and $m_h$ represent respectively the effective mass of a conduction electron and the effective mass of a hole. In general, $m_e$ and $m_h$ are not equal to $m$. The reason for their appearance is that in the modern theory of solids the electrons are described in terms of wave mechanics. As a result of the wave character of the electrons, the motion of an electron in a crystal under influence of an electric field is different from that of an electron in free space. The theory shows that this difference can be expressed formally by introducing an effective mass $m_e$ for an electron in a crystal so that the acceleration produced by an electric field $E$ is still given by the classical Newtonian equation $a = -eE/m_e$.

Assuming for the moment as an approximation that $m_e = m_h = m$, one finds from (6.7) upon substituting the numerical values

$$n \cong 5 \times 10^{21} T^{3/2} e^{-W_g/2kT} \text{ per m}^3 \tag{6.8}$$

where $T$ is expressed in degrees absolute. As an example, consider a semiconductor with an energy gap $W_g = 1$ ev, which is typical for semiconductors such as Ge and Si. At room temperature, $kT \cong 0.025$ ev, so that $W_g/2kT \cong 20$. This gives $n \cong 5 \times 10^{16}$ per m³, whereas the number of atoms is approximately $5 \times 10^{28}$ per m³. Hence, only a very small fraction of the valence electrons is actually excited into the conduction band at this temperature.

The exponent in (6.7) provides the predominant part of the temperature dependence of $n$, the $T^{3/2}$ factor varying relatively slowly. The functional relationship between $n$ and $T$ may be given in the form of a plot of $\log (n/T^{3/2})$ versus $1/T$; this yields a straight line in those cases for which the assumptions involved in the derivation of (6.7) are valid. A schematic plot of this kind is given in Fig. 6.7.

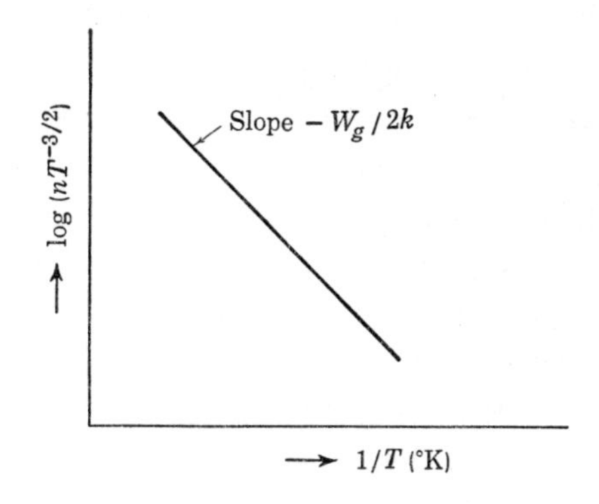


**Fig. 6.7.** Schematic illustration of the variation of the carrier density with temperature for an intrinsic semiconductor, according to equation (6.8).

Since the conductivity of a material is proportional to the density of charge carriers (see formula 5.13 for metals), the conductivity of a semiconductor depends strongly on the magnitude of the energy gap $W_g$. To illustrate this relationship, we consider in Table 6.3 the resistivity at room temperature and the energy gap for the elements in the fourth group of the periodic table. All these elements crystallize in the diamond structure. It is observed that $\rho$ increases very rapidly with increasing $W_g$ as a result

**Table 6.3.** Room temperature resistivity and energy gap of elements in the fourth group

| | C (diamond) | Si | Ge | Sn (gray) | Pb |
|---|---|---|---|---|---|
| $\rho$ (ohm m)....... | $10^{14}$ | 3000 | 0.47 | $2 \times 10^{-6}$ | $2 \times 10^{-7}$ |
| $W_g$ (ev)......... | 5.2 | 1.21 | 0.75 | 0.08 | No gap |

of the Boltzmann factor appearing in expression (6.7). Note the gradual transition from an insulator to a metal as one moves from diamond to lead; in lead, there is no energy gap and the density of electrons in the conduction band is essentially independent of temperature.

The considerations above apply not only to intrinsic elemental semiconductors, but also to intrinsic compound semiconductors, such as ZnO, PbS, etc.

## 6.4 The conductivity of intrinsic semiconductors

In Chapter 5 we saw that the conductivity of a metal is determined by the density of conduction electrons, and by their relaxation time; this led to the formula (5.13) $\sigma = ne^2\tau/m$. The conductivity of semiconductors can be treated on a similar basis, the difference being that, in general, one deals with *two* types of charge carriers, viz. electrons and holes. Thus, if there are $n_e$ conduction electrons and $n_h$ holes per m$^3$, we may write the conductivity of a semiconductor in the form

$$\sigma = \frac{n_e e^2 \tau_e}{m_e} + \frac{n_h e^2 \tau_h}{m_h} \tag{6.9}$$

Here, $\tau_e$ and $\tau_h$ are the *relaxation times* of the electrons and the holes respectively; $m_e$ and $m_h$ are the effective masses of electrons and holes. According to the results obtained in the preceding section, we have for an intrinsic semiconductor $n_e = n_h$. We can also define for a semiconductor two mobilities, one for electrons ($\mu_e$) and one for holes ($\mu_h$); these quantities are defined in analogy with (5.11) as follows:

$$\mu_e = e\tau_e/m_e \qquad \text{and} \qquad \mu_h = e\tau_h/m_h \tag{6.10}$$

They represent the average drift velocities produced by a field of 1 volt per m. Thus, for an intrinsic semiconductor with $n_e = n_h = n$ we may write

$$\sigma_{\text{intrinsic}} = ne(\mu_e + \mu_h) = ne^2\left(\frac{\tau_e}{m_e} + \frac{\tau_h}{m_h}\right) \tag{6.11}$$

where $n$ is given by expression (6.7). In an intrinsic semiconductor the charge carriers are scattered mainly by the lattice vibrations. On the basis of certain simplifying assumptions, the theory leads for this type of scattering to relaxation times which are proportional to $T^{-3/2}$. Assuming for the moment that this is correct, the theory thus predicts for the conductivity of an intrinsic semiconductor

$$\sigma_{\text{intrinsic}} = \text{constant} \times e^{-W_g/2kT} \tag{6.11a}$$

where we have used expression (6.7) for $n_e = n_h = n$. Although the relaxation times determined from experiment are not exactly proportional to $T^{-3/2}$, it is noted that the temperature dependence predicted by the theory is essentially determined by the Boltzmann factor, even if the constant in

(6.11a) varies somewhat with $T$. The form of (6.11a) is in agreement with experiment, as may be seen from Fig. 6.8, where the logarithm of $\sigma$ versus $1/T$ has been plotted for intrinsic Ge and Si. Note that such a curve permits one to evaluate the energy gap $W_g$ from the slope. By way of illustra-

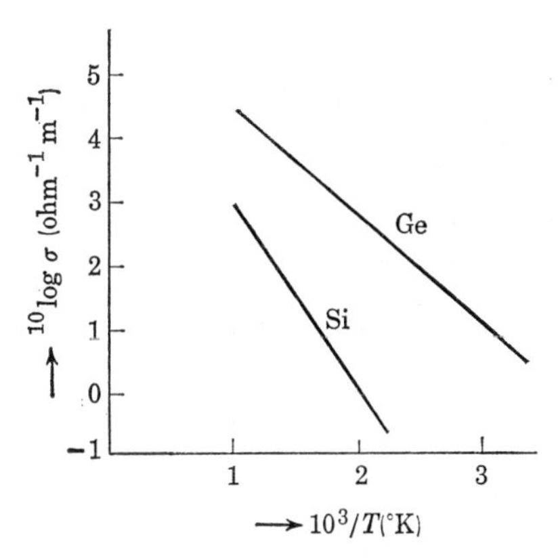


**Fig. 6.8.** The logarithm of the conductivity of germanium and silicon in the intrinsic region as a function of $1/T$.

tion we give here numerical values for the mobilities and carrier density in germanium, as derived from an analysis of the electrical properties of this material:

$$n_e = n_h = n = 2 \times 10^{22} T^{3/2} \times 10^{-3700/T}$$
$$\mu_e(T) = 3.5 \times 10^3 T^{-1.67} \text{ m}^2 \text{ volt}^{-1} \text{ sec}^{-1}$$
$$\mu_h(T) = 9.1 \times 10^4 T^{-2.3} \text{ m}^2 \text{ volt}^{-1} \text{ sec}^{-1}$$
$$\mu_e \text{ at room temperature} = 0.38 \text{ m}^2 \text{ volt}^{-1} \text{ sec}^{-1}$$
$$\mu_h \text{ at room temperature} = 0.18 \text{ m}^2 \text{ volt}^{-1} \text{ sec}^{-1}$$

For silicon, the mobilities as function of temperature are given by

$$\mu_e(T) = 4.0 \times 10^5 T^{-2.6} \text{ m}^2 \text{ volt}^{-1} \text{ sec}^{-1}$$
$$\mu_h(T) = 2.5 \times 10^4 T^{-2.3} \text{ m}^2 \text{ volt}^{-1} \text{ sec}^{-1}$$
$$\mu_e \text{ at room temperature} = 0.17 \text{ m}^2 \text{ volt}^{-1} \text{ sec}^{-1}$$
$$\mu_h \text{ at room temperature} = 0.035 \text{ m}^2 \text{ volt}^{-1} \text{ sec}^{-1}$$

All these values should be considered as approximate, since they have been derived on the basis of an analysis which involves models which are known not to be exact.

## 6.5 Carrier densities in *n*-type semiconductors

Most of the engineering applications of semiconductors involve semiconductors which have been "doped" intentionally with specific impurities, rather than intrinsic material. Thus, by adding a small amount of a particular impurity to the melt of germanium, crystals may be grown in which these impurities are incorporated. From X-ray diffraction data it has been established that in most cases the impurity atoms occupy lattice positions which in the pure material are occupied by germanium atoms; the same is true when the pure material is silicon. Of particular importance in the case of Si and Ge is doping with elements from the third and fifth columns of the periodic table; the most frequently used elements from the third column are B, Al, Ga, and In, and from the fifth: P, As, and Sb. Addition of a fraction of a percent of such elements may increase the conductivity by several powers of ten. Also, the temperature dependence of the conductivity is strongly affected by the impurity content. The reader is reminded here that these features are altogether different from those obtained by adding small amounts of impurities to metals. The strong dependence of the electrical properties of semiconductors on impurity content may be understood from the arguments given presently.

Consider a crystal of germanium in which a small fraction of the Ge atoms is replaced by, say, Sb atoms. An antimony atom contains five valence electrons, and since it occupies a normal Ge-lattice site, it requires only four valence electrons to form four electron-pair bonds with its neighbors. Thus, the Sb atom has one valence electron which is not being used in the chemical bonds with its neighbors. It is therefore not surprising that this "extra" electron is not as strongly bound as the other valence electrons. How strongly is the extra electron bound; i.e., how much energy would be required to set it free? An approximate numerical answer to this question may be obtained as follows. Suppose the electron is taken away from the antimony site; we are then left with an $Sb^+$ ion, carrying a charge $+e$. Consequently, if the electron remains near the antimony site, it does so because it is attracted by the field of the $Sb^+$ ion. The extra electron may thus be pictured as moving around a positive ion of charge $+e$, in a way similar to the motion of an electron in a hydrogen atom around a proton of charge $+e$ (see Fig. 6.9). There is, however, an important difference

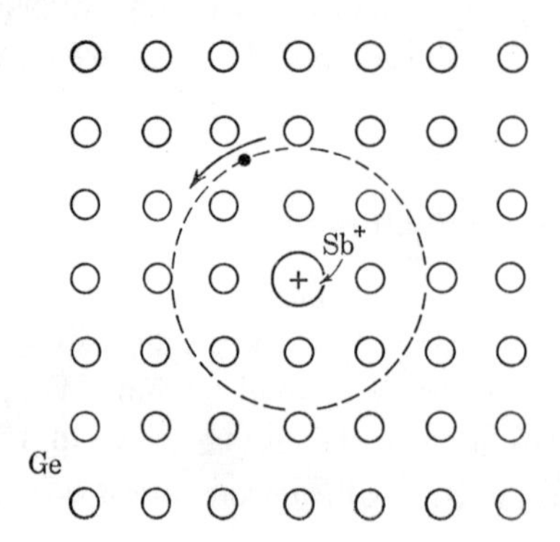


**Fig. 6.9.** Schematic representation of an antimony donor atom incorporated in a germanium lattice; the extra electron is pictured as revolving in a Bohr orbit around the $Sb^+$ ion.

between these two cases: the extra electron moving in the field of the $Sb^+$ ion sees the ion embedded in a dielectric medium, viz., germanium, which has a relative dielectric constant $\epsilon_r = 16$; the electron in a hydrogen atom, on the other hand, moves in free space. This difference is important, because the presence of the dielectric between the electron and the $Sb^+$ ion reduces the field strength and hence reduces the binding energy of the electron. This may be seen readily by following through the simple Bohr model for the present case. We assume the electron to move in a circular orbit of radius $r$; if $v$ represents the velocity of the electron in the orbit, we obtain the equilibrium condition

$$\frac{e^2}{4\pi\epsilon_0\epsilon_r r^2} = \frac{mv^2}{r} \tag{6.12}$$

The left-hand side represents the Coulomb force between the electron and the ion; the right-hand side is the centrifugal force. The total energy of the electron is then equal to

$$\begin{aligned} W &= W_{\text{kin}} + W_{\text{pot}} = \tfrac{1}{2}mv^2 + W_{\text{pot}} \\ &= \frac{e^2}{8\pi\epsilon_0\epsilon_r r} - \frac{e^2}{4\pi\epsilon_0\epsilon_r r} = -\frac{e^2}{8\pi\epsilon_0\epsilon_r r} \end{aligned} \tag{6.13}$$

Now, according to one of Bohr's quantum postulates, $2\pi$ times the angular momentum of the electron should be equal to an integer times Planck's constant $h$. If we consider the electron to be in its lowest energy state (the ground state), the integer is 1, so that

$$mvr = h/2\pi \tag{6.14}$$

Squaring both sides of this equation, and eliminating $v^2$ by means of (6.12), one obtains

$$r = \epsilon_0\epsilon_r h^2/me^2\pi \tag{6.15}$$

The energy of the electron may be obtained by substituting $r$ from (6.15) into (6.13), which gives

$$W = -me^4/8\epsilon_0^2\epsilon_r^2h^2 \tag{6.16}$$

From these results we may draw the following conclusions. According to (6.15), the radius of the electron orbit is proportional to $\epsilon_r$. Thus, according to this model the radius of the orbit of an electron moving in the field of an $Sb^+$ ion embedded in germanium is $\cong 16 \times 0.53 \cong 8.5$ angstroms. Also, according to (6.16), the energy of the electron is proportional to $1/\epsilon_r^2$. Since $-W$ represents the energy required to take the electron away from the Sb atom, and since the energy required to ionize a hydrogen atom is equal to $me^4/8\epsilon_0^2h^2 = 13.6$ ev, the ionization energy of the Sb atom is only approximately 0.05 ev. The model employed here would, of course, apply equally well to other elements of the fifth column, since only the dielectric constant $\epsilon_r$ appears as a parameter in the formulas. The model itself, however, is rather crude and it is not surprising, therefore, that the estimated value is not exactly equal to values obtained from an analysis of experimental data. Thus, in germanium, the ionization energies of P, As, and Sb are found to be approximately 0.01 ev, and differ somewhat from element to element.

The same reasoning applies as well to impurities from the fifth column in silicon. Since silicon has a somewhat lower dielectric constant ($\epsilon_r = 12$), one expects the ionization energies of these elements in Si to be somewhat higher than in germanium. This is indeed the case, as may be seen from the experimental values given here:

| Impurity | P | As | Sb |
|---|---|---|---|
| Ionization energy (ev) | 0.045 | 0.049 | 0.039 |

We should remark here that as the impurity content increases, the ionization energies decreases; the values given apply to low concentrations (say $\lesssim 10^{23}$ per m$^3$).

What are the consequences of the results obtained above? Let us first consider a crystal of germanium containing a small amount of Sb impurities at absolute zero. Since at $T = 0$, the electronic system is in its lowest energy state, all valence electrons will be in the valence band, and all the Sb impurities will be un-ionized. The energy band picture corresponding to this situation is depicted in Fig. 6.10(a). The picture is similar to that

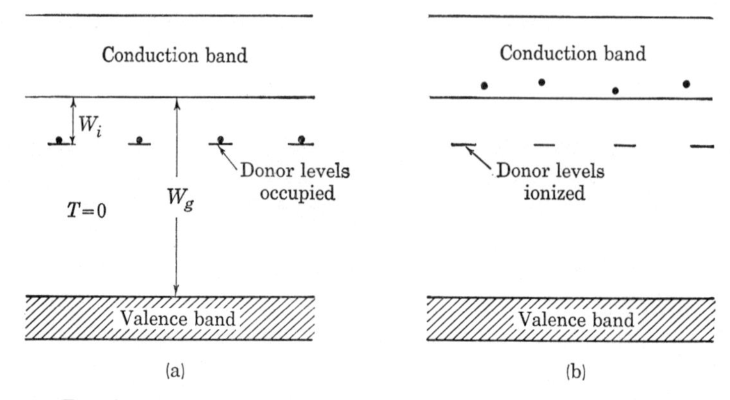


**Fig. 6.10.** In (a) the energy band scheme is drawn for a semiconductor containing donor atoms; at absolute zero, the donor atoms are not ionized and the donor electrons occupy levels $W_i$ below the bottom of the conduction band. In (b) the donor atoms are ionized by thermal excitation, leading to conduction by electrons.

for an intrinsic semiconductor in that the valence band is filled and the conduction band is empty. However, the "extra" electrons associated with the Sb atoms occupy states which lie only a fraction of a volt below the conduction band; in the case of Ge, the extra electron associated with an Sb atom lies 0.01 ev below the ionization continuum, which in the crystal corresponds to the conduction band. Thus, the short bars in Fig. 6.10(a) represent the energy levels for the extra electrons of the Sb atoms; the black dots represent the electrons themselves. As the temperature is raised, the lattice vibrations become more intense, and, by absorbing the proper amount of energy, some of the Sb atoms may become ionized; i.e., they release their electrons to the conduction band as indicated in Fig. 6.10(b). Since the ionization energy of the Sb atoms is much smaller than the energy $W_g$ required to break one of the valence bonds, the Sb atoms will donate electrons to the conduction band at much lower temperatures than the valence band will. The Sb atoms are called *donor levels*, because of their property of being able to donate electrons to the conduction band. The statistical mechanics involved in calculating the number of electrons donated to the conduction band at a certain temperature $T$ by a given number of impurity atoms is similar to that discussed in section 6.3. Thus, assuming for the moment that the temperature is still sufficiently low to neglect excitation from the valence band into the conduction band, we may proceed as follows: let there be $N_d$ donors, of which $n$ are ionized at a temperature $T$; there are then $(N_d - n)$ un-ionized donors. The rate at which electrons in the conduction band recombine with ionized donors is proportional to the number of conduction electrons and proportional to the number of ionized donors. Hence

$$(\partial n/\partial t)_{\text{recombination}} = -rn^2 \tag{6.17}$$

where $r$ is a constant. The rate at which electrons are excited from un-ionized donors into the conduction band is equal to

$$(\partial n/\partial t)_{\text{ionization}} = g(N_d - n) \tag{6.18}$$

where $g$ is a function of temperature. In equilibrium, the sum of (6.17) and (6.18) must be equal to zero. Hence,

$$n^2/(N_d - n) = \text{constant depending on } T \text{ alone} \tag{6.19}$$

It can be shown from statistical mechanics that the constant appearing in (6.19) is proportional to the Boltzmann factor exp $(-W_i/kT)$, where $W_i$ represents the ionization energy of the impurities. Thus,

$$n^2/(N_d - n) \propto e^{-W_i/kT} \tag{6.20}$$

Note that as $T$ increases, $n$ increases exponentially. When $kT$ becomes larger than $W_i$, nearly all donor levels are ionized, and $n$ approaches $N_d$. This is the case, for example, for Sb impurities ($W_i \cong 0.01$) at room temperature ($kT \cong 0.025$ ev). When the temperature is raised to sufficiently high values, there will also be excitation of electrons from the valence band to the conduction band. These results may be illustrated with reference to Fig. 6.11, where we have plotted schematically the density of electrons in

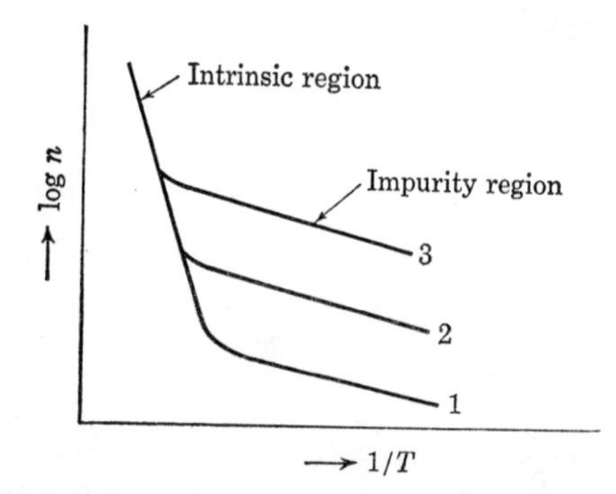


**Fig. 6.11.** Schematic representation of the logarithm of the electron concentration for a donor-doped semiconductor as a function of $1/T$. In the order, 1, 2, 3 the curves correspond to samples with increasing donor concentration.

a semiconductor containing various amounts of donor levels. At low temperatures, the density of conduction electrons is determined essentially by the donor concentration. The slope of the log $n$ versus $1/T$ curve is in that region determined by half the ionization energy [$n \ll N_d$ in (6.20)]. At high temperatures, one enters the intrinsic region, and because there are many more valence electrons than donor levels, the density of conduction electrons is ultimately determined only by excitation from the valence band. To give an example: for germanium nearly all impurities are ionized above liquid air temperatures; a sample doped so that its room temperature resistivity is 0.001 ohm m becomes intrinsic at approximately 200°C.

A semiconductor containing donor levels is referred to as *n-type* material, to indicate that the conduction below the intrinsic region is due to negative charge carriers. The carrier density is of course reflected in the conductivity of the material. The temperature dependence of the conductivity may be rather complicated as a result of the temperature dependence of the relaxation time and because the impurities themselves introduce new scattering processes. The details of these processes will not be discussed here.

## 6.6 p-type semiconductors

Returning to the elemental semiconductors, silicon and germanium, let us now consider the consequences of doping with elements from the third group of the periodic table; as an example, consider boron in germanium. The B atoms again occupy places in the lattice which in the pure material are occupied by Ge atoms. However, the boron atom has only three valence electrons, i.e. it is one electron short of being able to complete the four electron-pair bonds with its neighbors. The absence of an electron in one of these bonds is again called a hole, but in the present case the hole is "bound" to the boron atom, at least at absolute zero. If, as indicated in Fig. 6.12, an electron such as $A$ from a normal electron-pair bond would

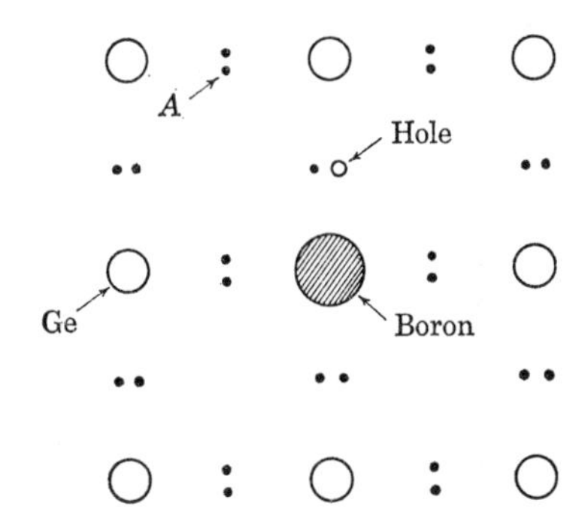


**Fig. 6.12.** Schematic illustration of a hole associated with incorporating a boron atom in a germanium lattice; the boron atom lacks one electron to form four electron-pair bonds. Actually, the hole may be pictured as revolving around the negative $B^-$ ion, in a way similar to the electron revolving around the $Sb^+$ ion in Fig. 6.9.

jump into the hole, subsequent jumps from other electrons into the hole would lead to hole conduction, as in the case of intrinsic semiconductors. However, some *energy is required for the first step in this process*, and it can be shown that the energy required for a valence electron to be excited into the hole is approximately equal to the ionization energy of donor levels in the same material; i.e., the bound holes of the boron atoms may be represented in the energy level diagram as occupying levels slightly above the top of the valence band, as indicated in Fig. 6.13(a). At absolute zero, the holes remain bound to the impurities, but as the temperature is raised, valence electrons may be excited into the bound holes, with the result that holes are created in the valence band, as indicated in Fig. 6.13(b). The bound hole levels are called *acceptor levels*, since they can accept electrons from the valence band. Materials of this kind are called *p-type*, because the conduction current is carried by holes. Much of the discussion of the preceding section with reference to electrons can be applied to the present case and it may therefore suffice to give an example of the conductivity as a function of temperature for a $p$-type semiconductor, viz., cuprous oxide. In Fig. 6.14 we have represented log $\sigma$ as a function of $1/T$ for $p$-type cuprous oxide, the concentration of acceptor levels being varied. At low temperatures, the conductivity results predominantly from holes produced by excitation of valence electrons into acceptor levels. The slope of the curve

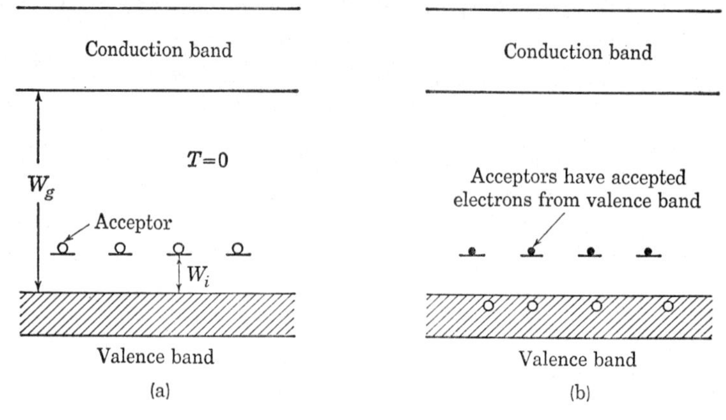


**Fig. 6.13.** At absolute zero, a hole such as the one indicated in Fig. 6.12 is bound to the acceptor atom, as indicated in (a) by the open circles above the valence band. At the expense of an energy $W_i$ electrons from the valence band may be raised into the acceptor levels; this is indicated in (b).

in that region is determined by the energy required to excite a valence electron into an acceptor level. As the temperature is raised, one ultimately enters the intrinsic region, where the slope is determined by the energy gap between the valence band and the conduction band. In the present material, the acceptor levels are formed in the following manner: In pure $Cu_2O$

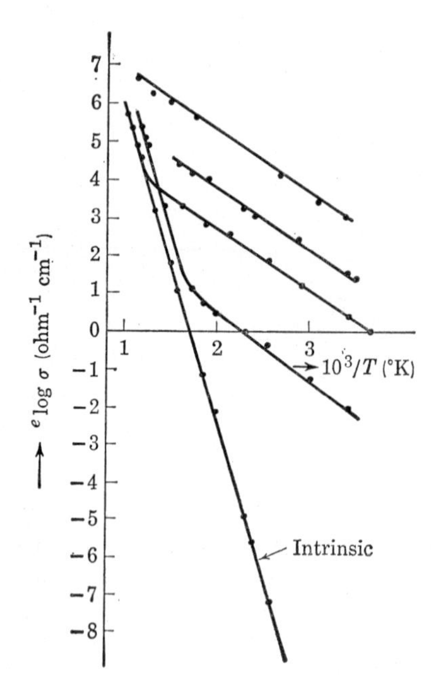


**Fig. 6.14.** The natural logarithm of the conductivity of various samples of cuprous oxide as function of $1/T$. The acceptor concentration increases as one goes to the higher lying curves. [After W. Juse and B. Kurtschatow, *Physik. Z. Sowjetunion* **2**, 453 (1933)]

there is one oxygen $O^{2-}$ ion for every two $Cu^+$ ions; one speaks in this case of a *stoichiometric composition.* If the material is heated in oxygen, the chemical formula is no longer exactly given by $Cu_2O$, because some "extra" oxygen is incorporated in the lattice. The extra oxygen atoms are apt to accept one or two electrons, because they have the tendency to become $O^{2-}$ rather than remain neutral. The oxygen atoms satisfy this need by accepting electrons from the valence band, which leads to mobile holes in the valence band. The acceptor levels in this case are thus the extra oxygen atoms incorporated in the lattice by heat treatment in oxygen. By varying the pressure of the oxygen, one may vary the acceptor concentration. As the acceptor concentration is increased, the conductivity of the material at low temperatures increases, as illustrated in Fig. 6.14.

## 6.7 Hall effect and carrier density

Since the conductivity of a material is determined by the product of the density and the mobility of charge carriers, a measurement of the conductivity alone does not permit one to evaluate these two factors separately. However, in case there is only one type of charge carrier, the density of the charge carriers can be found from a measurement of the so-called *Hall coefficient* $R_H$ of the material. In such cases then, measurement of $\sigma$ and of $R_H$ allows one to determine both the carrier density and mobility. It is from such measurements that the mobility data given in preceding sections were obtained. We shall now discuss the Hall effect and its interpretation, assuming the carriers are positive holes of charge $e$. In Fig. 6.15 consider a slab of $p$-type material in which there is a current density $J$ re-

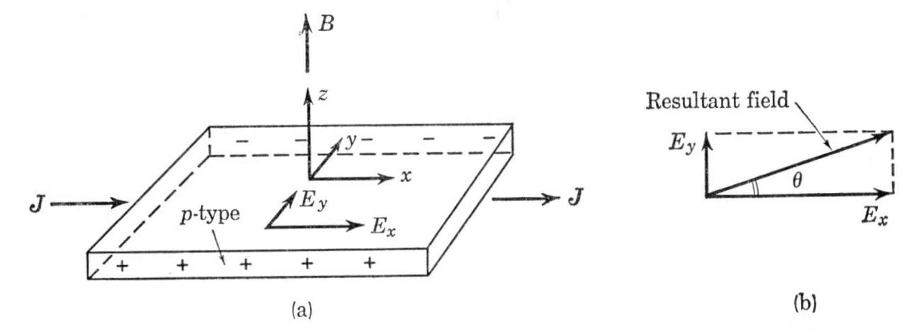


**Fig. 6.15.** Illustrating the geometry with reference to the discussion of the Hall effect for a $p$-type semiconductor. The voltage difference between the front and back faces of the slab in (a) gives the Hall voltage. In (b) the resultant field and the Hall angle are represented in the $xy$-plane.

sulting from an applied electric field $E_x$. Since the carriers are assumed to be positive, they will drift with an average velocity $\langle v_x \rangle$ in the $x$-direction. When a magnetic field of flux density $B_z$ (webers per m$^2$) is applied along the $z$-direction, the carriers will experience a Lorentz force perpendicular to $\langle v_x \rangle$ and to $B_z$; for the configuration of Fig. 6.15, the force will be directed along the negative $y$-axis. The magnitude of this force is given by $e\langle v_x \rangle B_z$. Thus, the positive carriers are driven toward the front face in the sample of Fig. 6.15, resulting in an excess of holes near the front face and a deficiency of holes near the back face. These charges will in turn create an electric field along the positive $y$-direction. In this way, then, an equilibrium situation is established in which the Lorentz force is just compensated by the force produced by the electric field along the $y$-axis. We may therefore write

$$eE_y = e\langle v_x \rangle B_z \tag{6.21}$$

On the other hand, the current density in the sample is given by

$$J_x = n_h e \langle v_x \rangle \tag{6.22}$$

where $n_h$ is the density of holes. Eliminating $\langle v_x \rangle$ from the last two equations we may define the Hall coefficient $R_H$ as

$$R_H = \frac{E_y}{B_z J_x} = \frac{1}{en_h} \tag{6.23}$$

The current density $J_x$ can be calculated from the total current and the cross section of the sample. The field $E_y$ may be found by attaching voltage probes at the front and back faces of the sample, and by dividing the measured Hall voltage by the width of the sample in the $y$-direction. Thus, $R_H$ can be obtained from measurements, and $n_h$ may be calculated. It should be noted here that proper averaging over the velocity distribution of the carriers gives $R_H = A/en_h$, where $A$ is a constant of the order of unity. For a simple semiconductor model, for example, one finds instead of (6.23)

$$R_H = \frac{3\pi}{8} \frac{1}{en_h} \tag{6.24}$$

The reader may verify for himself that if the carriers were electrons instead of holes, the polarity of the voltage measured across the Hall probes would be reversed. Hence, for electrons one finds in analogy with (6.24)

$$R_H = \frac{E_y}{B_z J_x} = -\frac{1}{en_e} \tag{6.25}$$

where $n_e$ represents the density of conduction electrons. We thus conclude that a measurement of the Hall coefficient of a semiconductor with one kind of charge carriers provides us with:

(a) the *sign of the charge carriers,*

(b) the *density of charge carriers.*

From the considerations above it follows that the total electric field in the sample of Fig. 6.15 makes a non-vanishing angle $\theta$ with the $x$-axis, as illustrated in Fig. 6.15(b). The angle $\theta$ is called the *Hall angle.*

When two types of charge carriers are present at the same time, the Hall effect is more complicated and a measurement of the Hall coefficient does not permit one immediately to calculate the carrier densities. Thus, under the same assumptions as were involved in the derivation of (6.23) one finds

$$R_H = \frac{E_y}{B_z J_x} = \frac{1}{e}\frac{(n_h\mu_h^2 - n_e\mu_e^2)}{(n_h\mu_h + n_e\mu_e)^2} \tag{6.26}$$

where $n_e$ and $n_h$ are the electron and hole densities, and $\mu_e$ and $\mu_h$ the carrier mobilities. Note that the Hall coefficient is positive if $n_h\mu_h^2 > n_e\mu_e^2$ and negative if $n_h\mu_h^2 < n_e\mu_e^2$.

As an example of measurements of the Hall coefficient of semiconductors, we reproduce in Fig. 6.16 a set of curves for $n$-type silicon, plotted as

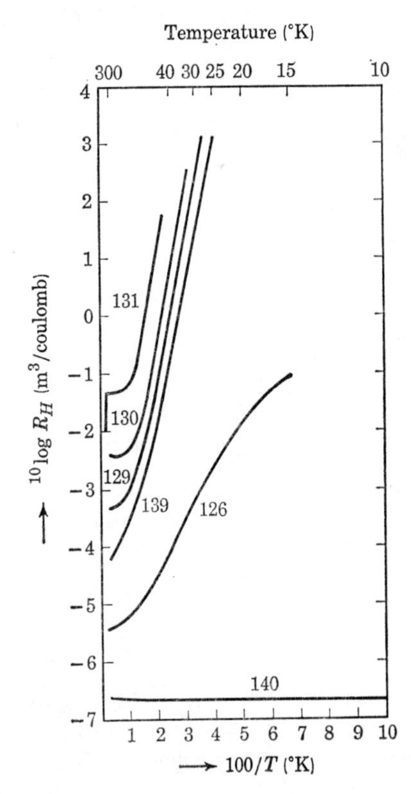


**Fig. 6.16.** Hall coefficient as function of temperature for single crystals of $n$-type silicon. The donor densities for these samples are approximately given by $7.4 \times 10^{18}/R_H$ per m³, where $R_H$ is the Hall coefficient at room temperature.

log $R_H$ versus $1/T$. Large Hall coefficients correspond to low carrier densities, and hence to purer samples. Sample 140 contains a considerable donor density, and, since the ionization energy of the donors decreases with increasing donor concentration, essentially all donors in this sample are ionized in the temperature region covered by the measurements; hence the constant value of $R_H$ for this sample. The other samples exhibit clearly the increase in carrier concentration with increasing temperature. From the straight portions of the upper curves, the donor ionization energy can be determined.

## References

W. C. Dunlap, *Introduction to Semiconductors*, Wiley, New York, 1957.

H. Y. Fan, "Valence Semiconductors, Germanium and Silicon," in *Solid State Physics* (F. Seitz and D. Turnbull, eds.), Academic Press, New York, 1955, vol. 1, pp. 284–367.

*Proc. IRE*, **43** (1955) (Solid State Electronics Issue).

"Semiconducting Materials," *Proc. Reading Conference*, Butterworths, London, 1951.

W. Shockley, *Electrons and Holes*, van Nostrand, New York, 1950.

D. A. Wright, *Semiconductors*, Wiley, New York, 1951.

## Problems

**6.1** Silicon and germanium have the diamond structure represented in Fig. 5.2. How many atoms are there on the average in a volume equal to $a^3$ where $a$ is the edge of the elementary cube in Fig. 5.2?

**6.2** Let $a$ represent the edge of the elementary cube shown in Fig. 5.2. For Si, $a = 5.43$ angstrom and for Ge, $a = 5.62$ angstrom. Calculate the number of atoms, $N$, per m³ in these materials.

**6.3** Let $n_e$ and $n_h$ represent the densities of free electrons and holes in a semiconductor. At a certain temperature $T$ let $g$ be the number of electrons excited thermally per unit volume from the valence band into the conduction band. Let the rate of recombination of electrons and holes be proportional to the product $n_e n_h$. If $n_i$ represents the carrier density for intrinsic material, show that in general $n_e n_h = n_i^2$ in thermal equilibrium.

**6.4** The resistivity of intrinsic germanium at 27°C is equal to 0.47 ohm m. Assuming electron and hole mobilities of respectively 0.38 and 0.18 m² volt⁻¹ sec⁻¹, calculate the intrinsic carrier density $n_i$ at 27°C.

**6.5** The resistivity of intrinsic silicon at 27°C is 3000 ohm m. Assuming electron and hole mobilities of respectively 0.17 and 0.035 $m^2$ $volt^{-1}$ $sec^{-1}$, calculate the intrinsic carrier density $n_i$ at 27°C.

**6.6** Germanium is doped with $1.00 \times 10^{-2}$ atomic percent of antimony. Assuming that at room temperature all antimony atoms are ionized, compute the electron and hole densities $n_e$ and $n_h$; you may assume that the electron density is determined only by the donors. (Hint: Make use of the result in problem 6.3.) From this information, calculate the resistivity of this material at room temperature if the electron and hole mobilities are 0.38 and 0.18 $m^2$ $volt^{-1}$ $sec^{-1}$, respectively.

**6.7** The Hall coefficient of a specimen of doped silicon is found to be $3.66 \times 10^{-4}$ $m^3$ $coulomb^{-1}$; the resistivity of the specimen is $8.93 \times 10^{-3}$ ohm m. Find the mobility and density of the charge carriers, assuming single carrier conduction.

**6.8** A specimen of a semiconductor is 1 mm thick and 1 cm wide. A magnetic flux density of 0.5 weber $m^{-2}$ is applied parallel to the 1 mm edge, and Hall voltage contacts are attached to measure the voltage across the width of the sample. The current flowing lengthwise through the sample is 10 milliamperes. If the Hall coefficient of the material is $3.66 \times 10^{-4}$ $m^3$ $coulomb^{-1}$, compute the voltage measured between the Hall contacts.

**6.9** A specimen of a semiconductor has a Hall coefficient of $3.66 \times 10^{-4}$ $m^3$ $coulomb^{-1}$ and a resistivity of $8.93 \times 10^{-3}$ ohm m. In a Hall effect experiment a magnetic flux density of 0.5 weber $m^{-2}$ is used. Find the Hall angle.

**6.10** Give a derivation of formula (5.26) for the Hall coefficient of a semiconductor with two types of carriers.

# 7

# *Junction Rectifiers and Transistors*

In this chapter we apply the information obtained in the preceding section to the processes occurring in *junction rectifiers* and *transistors*. Rectifying contacts occur not only between $n$- and $p$-type semiconductors, but also between metals and between metals and semiconductors. The main purpose of this chapter is to provide the reader with some insight into the physical processes which are responsible for the behavior of such devices, rather than to give a more or less complete review of the various kinds of rectifiers and transistors. For this reason, only the junction rectifier and transistor are discussed. The first three sections deal with preparatory material which is essential for an understanding of the physics of these devices.

## 7.1 Minority and majority carrier densities in semiconductors

In section 6.3 we discussed the density of conduction electrons and holes in an intrinsic semiconductor on the basis of the reaction

$$\text{valence electron} \leftrightarrows \text{conduction electron} + \text{hole} \tag{7.1}$$

Denoting the number of valence electrons per $m^3$ per second excited thermally into the conduction band by $g(T)$, the recombination coefficient by $r$, and the *intrinsic carrier concentration* by $n_i$, we have according to (7.1) for thermal equilibrium

$$g(T) = rn_i^2 \qquad \text{or} \qquad n_i^2 = g(T)/r \tag{7.2}$$

Thus, $n_i$ is determined only by the temperature of the material. We have also seen in sections 6.5 and 6.6 that a semiconductor may be doped with donors or acceptors, and in that case the density of one type of carrier predominates. Thus, in the presence of donor levels, the electrons are the *majority carriers* and the holes the *minority carriers*. In thermal equilibrium, let an impurity semiconductor contain $n_e$ electrons and $n_h$ holes per $m^3$ at a given temperature. It will be evident that even in this case, where $n_e \neq n_h \neq n_i$, reaction (7.1) must be in thermal equilibrium. Since neither the rate of generation $g(T)$ nor the recombination rate $r$ depend on the concentration of donors or acceptors for concentrations used in practical semiconductors, we must require that in thermal equilibrium

$$g(T) = rn_en_h \qquad \text{or} \qquad n_en_h = n_i^2 \tag{7.3}$$

Hence, $n_e$ and $n_h$ are related to the intrinsic carrier density. This provides a means for calculating the minority and majority carrier densities of a semiconductor, provided the conductivity of the material is known as well as $n_i$ and the mobility of the majority carriers. For example, let the conductivity of an $n$-type semiconductor be $\sigma$. Since the hole density is very small compared to the electron density (assuming the temperature is low enough that we are not in the intrinsic region), the conductivity is essentially determined by the electrons, i.e.

$$\sigma \cong n_e e \mu_e \tag{7.4}$$

where $\mu_e$ is the electron mobility. If the mobility is known, $n_e$ can be calculated from $\sigma$, and $n_h$ can be found from equation (7.3), provided $n_i$ is known.

In Table 7.1 we give values for $n_e$ and $n_h$ for samples of germanium of a given room-temperature conductivity, calculated in the manner outlined above. The conductivity range given here covers that of germanium crystals used in semiconductor rectifiers and transistors. For these calculations, the following values were used:

$$\mu_e = 0.36 \text{ m}^2 \text{ volt}^{-1} \text{ sec}^{-1}$$
$$\mu_p = 0.17 \text{ m}^2 \text{ volt}^{-1} \text{ sec}^{-1}$$
$$n_i = 2.5 \times 10^{13} \text{ m}^{-3}$$

**Table 7.1.** Majority and minority carrier concentrations for $n$- and $p$-type samples of germanium for a given conductivity $\sigma$ at room temperature

| $p$-Type Samples | | | $n$-Type Samples | | |
|---|---|---|---|---|---|
| $\sigma$ (ohm$^{-1}$ m$^{-1}$) | $n_e$ (m$^{-3}$) | $n_h$ (m$^{-3}$) | $\sigma$ (ohm$^{-1}$ m$^{-1}$) | $n_e$ (m$^{-3}$) | $n_h$ (m$^{-3}$) |
| $10^4$ | $1.70 \times 10^{15}$ | $3.68 \times 10^{23}$ | $10^4$ | $1.75 \times 10^{23}$ | $3.57 \times 10^{15}$ |
| $10^2$ | $1.70 \times 10^{17}$ | $3.68 \times 10^{21}$ | $10^2$ | $1.75 \times 10^{21}$ | $3.57 \times 10^{15}$ |
| 10 | $1.70 \times 10^{18}$ | $3.68 \times 10^{20}$ | 10 | $1.75 \times 10^{20}$ | $3.57 \times 10^{18}$ |

## 7.2 Drift currents and diffusion currents; the Einstein relation

In the preceding chapter we discussed the conduction mechanism in a semiconductor subjected to an electric field. As shown in section 6.4, the conductivity associated with the conduction electrons could be written in the form $\sigma_e = n_e e \mu_e$, where $\mu_e$ is the mobility of the electrons. The current density associated with the drift of the electrons due to the applied field would then be given by $J_e = n_e e \mu_e E$. However, an electron current may flow in a semiconductor even in the absence of an electric field, viz., if there exists a *gradient of the electron density*. Consequently, one deals in semiconductors frequently with two kinds of current: a *drift current* (due to the electric field) and a *diffusion current* (due to a gradient of the carrier concentration); these concepts apply, of course, to electrons as well as to holes. For our purpose it is sufficient to consider the case in which there exists a gradient of the carrier density in one direction only; we shall take this direction to be the $x$-direction. According to diffusion theory, the net flow of carriers through a plane of unit area perpendicular to the $x$-axis is proportional to $-dn/dx$, where $n$ is the density of carriers; the minus sign indicates the well-known fact that when the concentration increases along the positive $x$-direction, the flow of particles due to the diffusion process is in the negative $x$-direction. The proportionality constant relating the particle flow density and the concentration gradient is the *diffusion constant, D*. Thus, the hole current in any point $x$ may be written as the sum of two contributions, one from the field and one from the diffusion process:

$$J_h = n_h e \mu_h E - eD_h \frac{dn_h}{dx} \tag{7.5}$$

where $D_h$ is the diffusion constant of the holes. The charge appears in the diffusion contribution because the electric current is equal to the *particle* current times the charge $e$ per particle. For electrons, we may write down a similar expression for the current density:

$$J_e = n_e e \mu_e E + eD_e \frac{dn_e}{dx} \tag{7.6}$$

In this case the diffusion contribution is positive because the charge per electron is $-e$. The total current is given by the sum of equations (7.5) and (7.6).

There exists an important relationship between the diffusion coefficient and the mobility of the carriers, which is known as the *Einstein relation*. This relationship may be derived by considering a situation in a semiconductor in which there exists an electric field $E$ and a concentration gradient such that the resultant current is zero. Under these conditions of no current flow, the system is in thermal equilibrium and Boltzmann statistics applies. Consider, for example, a potential $V(x)$, producing in the point $x$ an electric field given by

$$E(x) = -\frac{dV}{dx} \tag{7.7}$$

In thermal equilibrium, the density of holes as function of $x$ is given by the Boltzmann expression

$$n_h(x) = Ce^{-eV/kT} \tag{7.8}$$

where $C$ is a constant. Hence, there will be a gradient of the hole density given by

$$\frac{dn_h}{dx} = -\frac{e}{kT} n_h \frac{dV}{dx} = \frac{e}{kT} n_h E(x) \tag{7.9}$$

Since the hole current vanishes in thermal equilibrium, we have in accordance with (7.5) and (7.9)

$$0 = n_h e \mu_h E - \frac{e^2}{kT} D_h n_h E$$

from which we obtain the Einstein relation

$$D_h = (kT/e)\mu_h \tag{7.10}$$

Hence, the *diffusion constant is proportional to the mobility* and for a given temperature may be calculated from the latter. A similar relationship holds, of course, for electrons:

$$D_e = (kT/e)\mu_e \tag{7.11}$$

For germanium we find, for example, from the room temperature mobilities given in connection with Table 7.1

$$D_h = 0.0044 \text{ m}^2 \text{ sec}^{-2}$$

$$D_e = 0.0093 \text{ m}^2 \text{ sec}^{-2}$$

In subsequent sections we shall see that diffusion currents play an important role in rectifier and transistor action.

## 7.3 The continuity equation for minority carriers

An important equation which governs the behavior of minority carriers in rectifying junctions and transistors is the *continuity equation;* the quantitative theory of junctions and transistors is, in fact, based on this equation. This equation will be derived in this section for the minority carriers (electrons) in a $p$-type semiconductor. Consider then a $p$-type semiconductor which for the moment will be assumed to carry no electron current or hole current. Let $n_{e0}$ and $n_{h0}$ represent the thermal *equilibrium densities* of minority and majority carriers respectively. In accordance with (7.3) we may then write

$$n_{e0}n_{h0} = n_i^2 = g/r \tag{7.12}$$

Suppose now that in some way or other the electron concentration at a given instant $t = 0$ is different from $n_{e0}$, say

$$n_e = n_{e0} + \delta n_e \tag{7.13}$$

It can be shown that for semiconductors with conductivities of the kind used in transistors and rectifiers, the hole density must also be different from the equilibrium concentration $n_{h0}$, so as to maintain *charge neutrality.* Thus, the hole density may be written as

$$n_h = n_{h0} + \delta n_h \tag{7.14}$$

where $\delta n_h \cong \delta n_e$. If the semiconductor has an excess density of electrons $\delta n_e$ at the instant $t = 0$, what will the electron density be for times when $t > 0$, assuming the semiconductor is left to itself? This question may be answered in words in the following manner:

$$\begin{bmatrix}\text{time rate}\\ \text{of increase}\\ \text{of electrons}\end{bmatrix} = \begin{bmatrix}\text{rate of}\\ \text{thermal generation}\\ \text{of electrons}\end{bmatrix} + \begin{bmatrix}\text{rate of}\\ \text{increase due to}\\ \text{recombination}\end{bmatrix} \tag{7.15}$$

Mathematically, this may be written as

$$\left(\frac{dn_e}{dt}\right)_{J_e=0} = g - r(n_{e0} + \delta n_e)(n_{h0} + \delta n_h) \tag{7.16}$$

We have included the condition $J_e = 0$ to remind us of the fact that to this point we have assumed that there is no current in the semiconductor. Writing out the last term in (7.16) we find

$$r(n_{e0} + \delta n_e)(n_{h0} + \delta n_h) = r[n_{e0}n_{h0} + n_{h0}\delta n_e + n_{e0}\delta n_h + \delta n_e \delta n_h] \tag{7.17}$$

At this point we make use of the fact that we are discussing the *minority carriers;* i.e., since we are dealing with a $p$-type semiconductor we may take $n_{h0} \gg n_{e0}$. Since $\delta n_e \cong \delta n_p$ we may then, to a first approximation, neglect the last two terms in (7.17). Realizing further that $rn_{e0}n_{h0} = g$, in accordance with (7.12), we find from (7.16)

$$\left(\frac{dn_e}{dt}\right)_{J_e=0} = -rn_{h0}\delta n_e \tag{7.18}$$

Now, since $n_e = n_{e0} + \delta n_e$, where $n_{e0}$ is the time-independent equilibrium density, it follows that

$$\frac{d}{dt}(\delta n_e) = -rn_{h0}\delta n_e = -\frac{\delta n_e}{\tau_e} \tag{7.19}$$

Here, $\tau_e$ is called the *lifetime* of the electrons; note that $\tau_e$ is determined only by the recombination coefficient $r$ and by the equilibrium hole density; i.e., $\tau_e$ is a constant if the approximations made above are justified. The physical meaning of $\tau_e$ becomes clearer by integrating (7.19):

$$(\delta n_e)_t = (\delta n_e)_{t=0}\, e^{-t/\tau_e} \tag{7.20}$$

This equation states that the *excess electron density will disappear exponentially with time,* the time constant being the lifetime $\tau_e$ of the electrons.

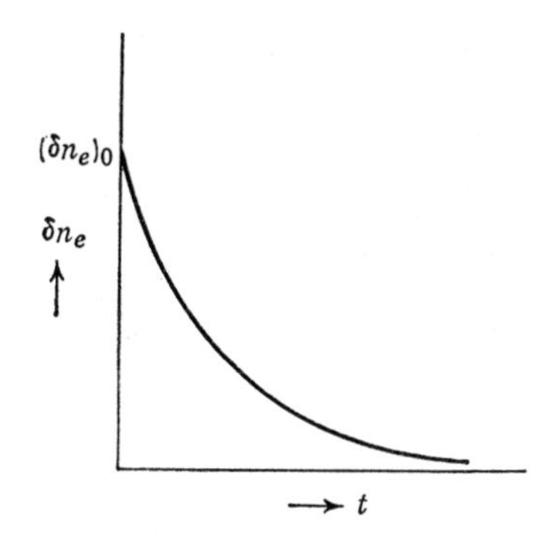


**Fig. 7.1.** Schematic representation of the decay of the excess of minority carriers according to formula (7.20).

This result is illustrated in Fig. 7.1. For later use, it is convenient to write our result in the form

$$\left(\frac{dn_e}{dt}\right)_{J_e=0} = \frac{n_{e0} - n_e}{\tau_e} \tag{7.21}$$

which follows immediately from (7.18) and (7.19).

What equation will the minority carriers satisfy in case there is an electron current flowing as well? In words we may write the answer to this question in the following form:

$$\begin{bmatrix}\text{total rate of}\\ \text{change of } n_e\end{bmatrix} = \begin{bmatrix}\text{rate of change}\\ \text{of } n_e \text{ for } J_e = 0\end{bmatrix} + \begin{bmatrix}\text{rate of change}\\ \text{of } n_e \text{ due to } J_e\end{bmatrix} \tag{7.22}$$

The first term on the right-hand side is given by (7.21), which leaves us with the problem of finding a mathematical expression for the last term. With reference to Fig. 7.2 consider two planes of unit area at $x$ and $x + dx$. If the current density $J_e$ is independent of $x$, there are presumably just as many particles entering the volume element between $x$ and $x + dx$ through the plane at $x$, as there are particles leaving through the plane at $x + dx$. Hence, if $J_e$ is independent of $x$, the last term in (7.22) is zero. Consider

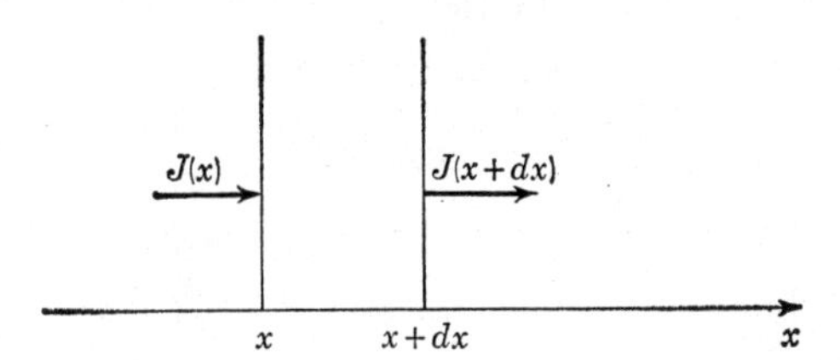


**Fig. 7.2.** Illustrating two planes perpendicular to the $x$-axis at the points $x$ and $x + dx$. The current density at $x$ is $J(x)$; the current density at $x + dx$ is $J(x + dx) = J(x) + (dJ/dx)\,dx$.

now the case where $J_e = J_e(x)$, so that the derivative $dJ_e/dx$ is *not* equal to zero. The *particle* current density $J_p$ is equal to $-(1/e)J_e$, where $-e$ is the electronic charge. The number of particles per m² per second entering the volume element at $x$ is equal to $J_p(x)$; the number of particles per m² per second leaving the volume element at $x + dx$ is equal to $J_p(x + dx)$. Hence, the *net* increase of the number of particles in the volume element $dx$ due to the current is

$$\left(\frac{\partial n_e}{\partial t}\right)_{\text{current}} \cdot dx = J_p(x) - J_p(x + dx) = -\frac{dJ_p}{dx}\,dx = \frac{1}{e}\frac{dJ_e}{dx}\,dx \tag{7.23}$$

Hence, the last term of (7.22) is equal to $-(1/e)(dJ_e/dx)$; i.e., this term depends on the *derivative of the current* with respect to $x$. Substituting (7.23) and (7.21) into (7.22) we then obtain the continuity equation which must be satisfied by the minority carriers:

$$\frac{dn_e}{dt} = \frac{n_{e0} - n_e}{\tau_e} + \frac{1}{e}\frac{dJ_e}{dx} \tag{7.24}$$

For the holes as minority carriers in an $n$-type semiconductor one finds the corresponding equation to be

$$\frac{dn_h}{dt} = \frac{n_{h0} - n_h}{\tau_h} - \frac{1}{e}\frac{dJ_h}{dx} \tag{7.25}$$

The current densities $J_e$ and $J_h$ are given by equations (7.6) and (7.5); *these four equations are the basic equations underlying the theory of junction rectifiers and transistors.*

## 7.4 Semi-quantitative discussion of the *n-p* junction rectifier

In this section we shall consider the current-voltage relationship for a junction consisting of an $n$-type semiconductor and a $p$-type semiconductor, as indicated schematically in Fig. 7.3(a). The $n$-type region contains donor levels and thus has a large concentration of electrons and a low concentration of holes; the $p$-type region contains acceptor levels, giving it a large concentration of holes and a low concentration of electrons, as indi-

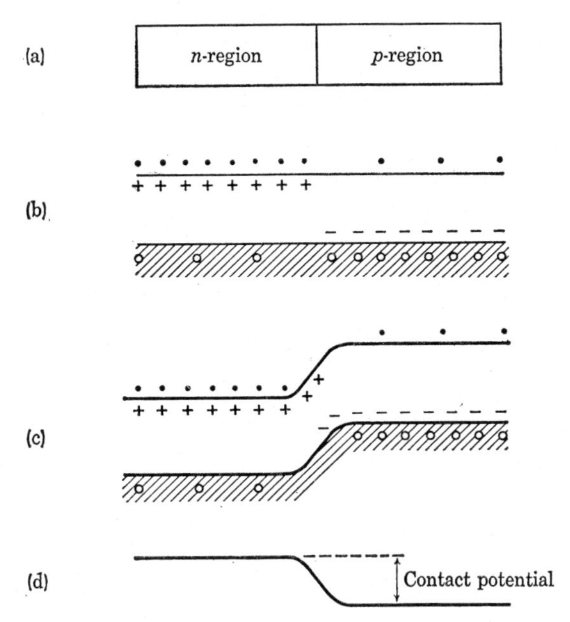


**Fig. 7.3.** (a) represents the $n$- and $p$-regions of the junction. (b) represents the electrons as black dots in the conduction band and the holes as open circles in the valence band for the two regions; this situation does *not* correspond to equilibrium. The ionized donors are represented by positive signs, the ionized acceptors by negative signs. In (c) the junction is in thermal equilibrium; note the positive space charge on the $n$-side and the negative space charge on the $p$-side of the junction. The potential distribution in (c) corresponds to electrons. In (d) the potential distribution is drawn for holes.

cated in Fig. 7.3(b). For simplicity we shall assume for the moment that the change from $n$- to $p$-type is abrupt. In practice this corresponds to a *fused junction* (*alloy-junction*). In so-called *grown junctions*, there is a more gradual transition from the $n$-type to the $p$-type semiconductor.

The situation drawn in Fig. 7.3(b) is unstable, and it will be evident that under influence of the concentration gradients near the junction, electrons will diffuse from $n$ to $p$ and holes will diffuse from $p$ to $n$. Let us consider the consequences of these processes. As we take electrons away from the $n$-region, we are left with positive ionized donors which are not compensated any longer by the negative electronic charge; thus, in the $n$-region a positive *space charge* will be set up. Consequently, the potential energy of an electron will become a function of the coordinate $x$. Similarly, the $p$-region will acquire a negative space charge near the junction, because negative acceptor levels are no longer compensated by holes. This will give rise to an energy band picture as depicted in Fig. 7.3(c); the reader is reminded of the fact that the potentials refer to electrons, i.e. to negative particles. In equilibrium, the diffusion of electrons uphill is just compensated by the electrons coming downhill as a result of the field set up by the space charge. The same holds for the holes, but for the moment we shall concentrate our attention on the electrons. It is evident that in the equilibrium state, the potential difference between the bulk $n$-region and between the bulk $p$-region must be of the order of the gap width divided by $e$. The potential difference set up in this manner is called the *contact potential* of the junction. For the usual germanium junctions this contact potential amounts to approximately half a volt. The potential as seen by the holes in the configuration corresponding to Fig. 7.3(c) is represented in Fig. 7.3(d); i.e., holes wishing to go from $p$ to $n$ must climb the same potential hill as electrons in going from $n$ to $p$.

In the equilibrium situation depicted in Fig. 7.3(c) let $J_{e0}$ represent the

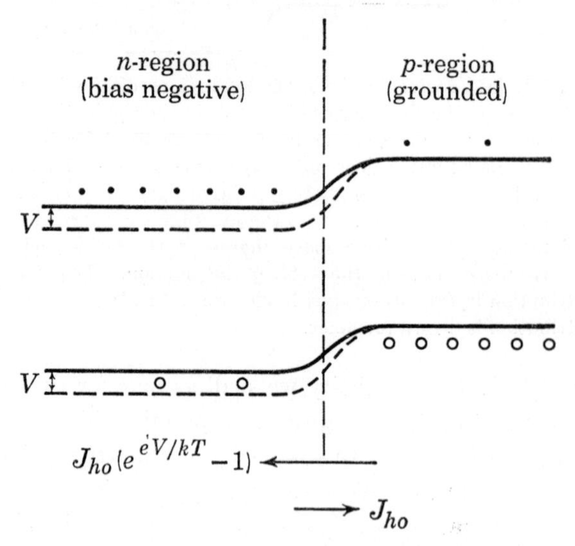


**Fig. 7.4.** Energy band picture for a forward biased $n$-$p$ junction; the $n$-region is made $V$ volts negative relative to the $p$-region. The dashed curves indicate the bands in the absence of bias. The hole currents are indicated.

magnitude of the two compensating electron currents, one associated with electrons moving uphill, the other with electrons moving downhill. Suppose now that the junction is *biased* so as to make the $n$-region negative with respect to the $p$-region. Assuming the latter is connected to ground, the potential seen by the electrons is then as indicated in Fig. 7.4. The electron current associated with the motion of electrons from $p$ to $n$ is not affected by the bias. However, the electrons tending to move from $n$ to $p$ now see a smaller potential hill than before; in fact, the probability that an electron will climb the hill is increased by a factor of $e^{eV/kT}$, where $V$ is the applied potential between the $p$-region and the $n$-region. Consequently, for the biased junction there will be a net electron current given by

$$J_e = J_{e0}(e^{eV/kT} - 1) \tag{7.26}$$

A similar reasoning may be applied to the hole current; the holes moving from $p$ to $n$ also see a lower potential barrier than in the unbiased situation. Thus, the biased junction carries a net hole current

$$J_h = J_{h0}(e^{eV/kT} - 1) \tag{7.27}$$

Consequently, the total current through the junction as given by the sum of (7.26) and (7.27) is equal to

$$J = (J_{e0} + J_{h0})(e^{eV/kT} - 1) \tag{7.28}$$

Since $(J_{e0} + J_{h0})$ is a constant, the current density increases exponentially with the voltage when applied so as to make the $n$-region negative with respect to the $p$-region; this type of bias is called *forward bias*, since it gives a much larger total current than a bias of the same magnitude applied in the opposite direction. In fact, if the $n$-region is biased positively relative to the $p$-region, we find from similar reasoning as above

$$J = (J_{e0} + J_{h0})(1 - e^{-eV/kT}) \tag{7.29}$$

where $V$ refers now to the *reverse bias*. Note that the reverse current approaches a saturation value $(J_{e0} + J_{h0})$ as the reverse bias is increased.

In Fig. 7.5 we have represented the forward and reverse currents characteristics for a $p$-$n$ junction; there is evidently good agreement between the theoretical curves (7.28) and (7.29) and measured points. It is observed that the system under discussion has a *non-linear* current-voltage characteristic and that it acts as a *rectifier*.

The simplified treatment presented here, of course, gives no information concerning the magnitude of the equilibrium currents $J_{e0}$ and $J_{h0}$. In order to calculate these currents in terms of the properties of the material, it is

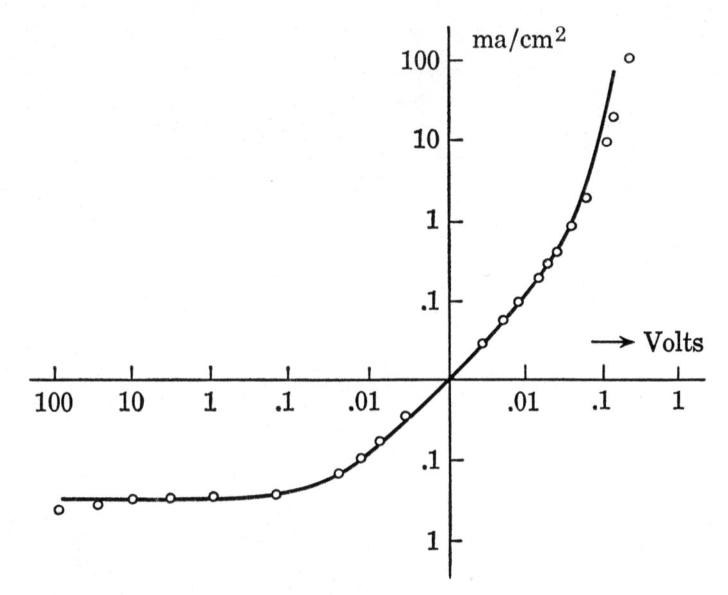


**Fig. 7.5.** Current-voltage characteristic of a *p-n* junction. The dots are measured points, the curve is theoretical. [After W. Shockley, *Proc. IRE* **40**, 1289 (1952)]

necessary to make a study of the continuity equation derived in the preceding section; this will be done in the next section.

## 7.5 Quantitative treatment of the *n-p* junction rectifier

Although the form of the equations for the forward and reverse currents derived in the preceding section are essentially correct, the derivation does not give much insight into the actual mechanism. In order to see in more detail what is actually happening in and near the junction, we refer to Fig. 7.6, where we have represented the energy band picture corresponding to a forward biased *n-p* junction, the potentials referring to electrons rather than to holes. We shall assume that for $x < 0$, and for $x > x_p$, the bottom of the conduction band is essentially horizontal. Actually, because of the current flow, there will be a slight gradient of the potential in the bulk *n*- and *p*-regions, but since the conductivity of the material is relatively high, the potential drop across these regions will be small compared to that across the barrier between them.

We shall see below that in order to discuss the mechanism of conduction in detail, it is necessary to know the electron and hole concentrations at the points $x = 0$ and $x = x_p$. These concentrations constitute the *boundary*

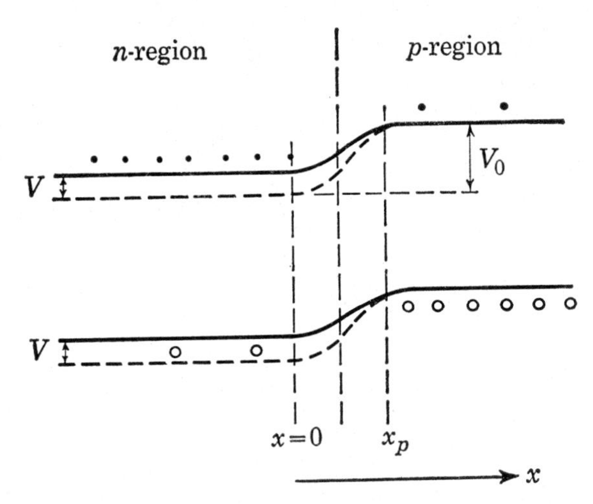


**Fig. 7.6.** Forward biased $n$-$p$ junction. The dashed curves give the energy bands in thermal equilibrium, corresponding to the contact potential $V_0$. The forward bias $V$ is indicated.

*conditions* which will be used to solve the continuity equations for minority carriers in the bulk $n$- and $p$-regions. We shall express these concentrations in terms of the following quantities:

$n_{e0}$ = density of electrons in *p-type* material in thermal equilibrium
$n_{h0}$ = density of holes in *n-type* material in thermal equilibrium
$V_0$ = voltage difference in absence of current between $p$-region and $n$-region
$V$ = applied forward bias

First consider the case when $V = 0$; i.e., the junction is in thermal equilibrium. The electron concentration for the region $x \leqslant 0$ is then given by

$$n_e = n_{e0}e^{eV_0/kT} \tag{7.30}$$

This follows from Boltzmann statistics and from the fact that the concentration of electrons for $x \geqslant x_p$ is equal to $n_{e0}$. What happens now when the forward voltage $V$ is applied? Holes are then injected from the $p$-region into the $n$-region, so that the *hole concentration near* $x = 0$ *will be larger than* $n_{h0}$. Since the material should remain neutral, the electron concentration at $x = 0$ must increase somewhat over that given by (7.30). However, the increase in the electron density at $x = 0$ required to satisfy the neutrality condition is very small compared to (7.30). Thus, we may assume that (7.30) gives the electron concentration in the region $x \leqslant 0$ even if current flows. If that is so, we can find the concentration of electrons in the point $x = x_p$ in the case of a forward bias from (7.30) by applying

Boltzmann statistics.* We thus find

$$n_e(x_p) = n_e(0)e^{-e(V_0-V)/kT} = n_{e0}e^{eV/kT} \tag{7.31}$$

Note that as a result of the forward bias, the electron density at $x_p$ is higher than it was in the absence of the bias, by a factor $e^{eV/kT}$. This illustrates clearly the *injection* of electrons into the $p$-region by the $n$-region. For large distances into the $p$-region, the electron density is, of course, equal to the thermal equilibrium value $n_{e0}$.

Without repeating the arguments here, the reader may readily verify for himself that the hole density at the point $x = 0$ is given by

$$n_h(0) = n_{h0}e^{eV/kT} \tag{7.32}$$

which expresses the injection of holes into the $n$-region.

We shall now proceed to calculate the total current flowing across the junction. For the moment, let us consider the electronic current in the region $x > x_p$, i.e. the minority carrier current in the $p$-region. In this region, the continuity equation (7.24) applies:

$$\frac{dn_e}{dt} = \frac{n_{e0} - n_e}{\tau_e} + \frac{1}{e}\frac{dJ_e}{dx} \tag{7.33}$$

where $n_e$ is the electron density as function of $x$ in the region $x > x_p$. The electron current density $J_e$ is in general given by expression (7.6). However, since the field strength in this region is negligible, we only have to consider the diffusion current, i.e.

$$J_e = eD_e\frac{dn_e}{dx} \tag{7.34}$$

In the state of steady flow of current, $dn_e/dt = 0$ anywhere. Thus, the equation describing the behavior of the electrons for purposes of the present problem is obtained by substituting (7.34) into (7.33) and by putting $dn_e/dt = 0$, i.e.

$$n_e(x) = n_{e0} + D_e\tau_e\frac{d^2n_e}{dx^2} = n_{e0} + L_e^2\frac{d^2n_e}{dx^2} \tag{7.35}$$

The quantity $L_e = (D_e\tau_e)^{1/2}$ has the dimensions of length, and is called the *diffusion length:* the reason for this name will become evident later. Since

* Strictly speaking, Boltzmann statistics is valid only *in thermal equilibrium;* i.e., if in expression (7.6) $J_e = 0$. However, for the material used in transistors, it turns out that each of the two terms on the right-hand side of (7.6) is of the order of $10^8$ ampere m$^{-2}$; this estimate is based on a barrier thickness $x_p \approx 10^{-6}$ m. The actual currents flowing under forward bias in the junction, on the other hand are only of the order of 100 ampere m$^{-2}$, which is negligibly small compared to the two terms on the right-hand side of (7.6). Consequently, Boltzmann statistics may be used even if current flows.

$n_{e0}$ is a constant, (7.35) may be written as

$$n_e(x) - n_{e0} = L_e^2 \frac{d^2(n_e - n_{e0})}{dx^2} \tag{7.36}$$

The general solution of this equation is of the form

$$n_e(x) - n_{e0} = Ae^{-x/L_e} + Be^{x/L_e} \tag{7.37}$$

where $A$ and $B$ are two constants which are determined by the boundary conditions. One of these boundary conditions is that for very large values of $x$, $n_e(x)$ should become equal to the equilibrium minority carrier density $n_{e0}$. Consequently, $B = 0$, because otherwise $n_e$ would go to infinity for $x \to \infty$. The constant $A$ is determined by the boundary condition (7.31); this gives

$$Ae^{-x_p/L_e} = n_{e0}(e^{eV/kT} - 1)$$

Calculating $A$, we then find from (7.37) the solution

$$n_e(x) - n_{e0} = n_{e0}e^{(x_p - x)/L_e}(e^{eV/kT} - 1) \quad \text{for } x \geqslant x_p \tag{7.38}$$

For a given bias voltage $V$, we see that the electron density decreases exponentially to $n_{e0}$ for increasing values of $x$ beyond $x_p$, as illustrated in Fig. 7.7. Note that $L_e$ is a measure for the length into the $p$-type material

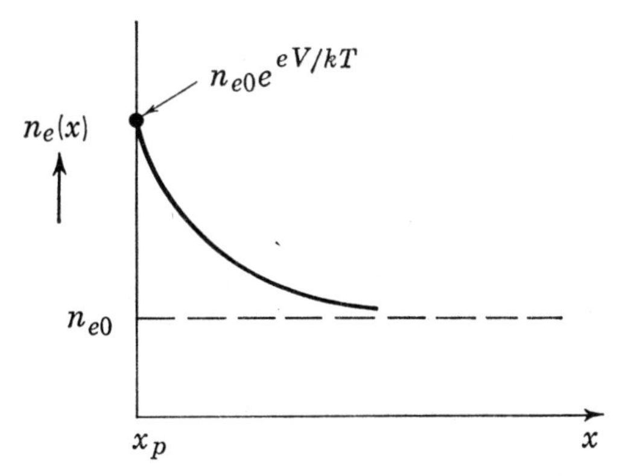


**Fig. 7.7.** The electron concentration in the $p$-region decreases with increasing distance from the value $n_{e0}e^{(eV/kT)}$ at $x_p$ to $n_{e0}$ for large values of $x$, in accordance with equation (7.38).

over which the excess density has been reduced by a factor $1/e$ (where $e = 2.7$). The reason for this decrease is that at $x_p$ the electron density is higher than the equilibrium density $n_{e0}$ as a result of electron injection from the $n$-region. This produces a diffusion of excess electrons into the $p$-region, but at the same time the density of excess electrons gradually disappears as one moves into the $p$-region because of recombination with holes. The depth over which the excess electrons are "noticeable" is of the order of $L_e$, hence the name "diffusion length" for $L_e$.

Since we have now an expression for $n_e(x)$, we may calculate the electron current from (7.34). Differentiating (7.38) with respect to $x$, substituting the result into (7.34) and finally putting $x = x_p$, we find for the electron current density at $x_p$

$$J_e(x_p) = -\frac{eD_e n_{e0}}{L_e}(e^{eV/kT} - 1) \tag{7.39}$$

The minus sign indicates that the conventional positive current flows from $p$ to $n$.

The hole current density in the $n$-region at $x = 0$ may be obtained in a similar way. It may suffice to give the result:

$$J_h(0) = -\frac{eD_h n_{h0}}{L_h}(e^{eV/kT} - 1) \tag{7.40}$$

Here, $L_h = (D_h \tau_h)^{1/2}$ is the diffusion length of holes in the $n$-region, and again the minus sign indicates that the current flows from $p$ to $n$. Thus, the total current density flowing across the junction is given by the sum of (7.39) and (7.40); the magnitude of the total current density flowing from $p$ to $n$ is then

$$J = e\left(\frac{D_e n_{e0}}{L_e} + \frac{D_h n_{h0}}{L_h}\right)(e^{eV/kT} - 1) \tag{7.41}$$

We note that the form of this equation is the same as that of expression (7.28) given in the preceding section.

It is of importance for the discussion of transistors later, to say a few words about the ratio of the electron and hole currents. It follows from (7.40) and (7.41) that $J_e/J_h = (D_e n_{e0} L_h)/(D_h n_{h0} L_e)$. As a consequence of the Einstein relation between the diffusion coefficient and mobility of carriers, we may write $D_e/D_h = \mu_e/\mu_h$. Furthermore, if $n_{hp}$ represents the hole density in the $p$-region *in thermal equilibrium*, it must follow according to (7.3) that $n_{e0} n_{hp} = n_i^2$. Similarly, if $n_{en}$ represents the equilibrium density of electrons in the $n$-region we must have $n_{en} n_{h0} = n_i^2$. Hence

$$\frac{J_e}{J_h} = \frac{\mu_e n_{en} L_h}{\mu_h n_{hp} L_e} = \left(\frac{\sigma_n}{\sigma_p}\right)\left(\frac{L_h}{L_e}\right) \tag{7.42}$$

Apart from the ratio $L_h/L_e$ which is of the order of unity, the ratio of the electron to hole current is determined by the conductivity $\sigma_n$ of the $n$-region over the conductivity $\sigma_p$ of the $p$-region. Thus, if the conductivity of the $n$-region is 100 times as large as the conductivity of the $p$-region, the current across the junction is carried for 99 percent by electrons and for 1 percent by holes. Typical values for $L_e$ and $L_h$ are of the order of $10^{-3}$ m.

For reverse bias, one obtains a current through the junction which is of the same form as equation (7.29), the identification of $J_{e0}$ and $J_{h0}$ being the same as above.

The procedure followed here in the calculation of the junction current may be summarized as follows:

(a) set up the boundary conditions for the minority carriers in the $n$- and $p$-regions for $x = x_p$ and $x = 0$;

(b) write down the continuity equation for the minority carriers for the region $x \geqslant x_p$. Solving the equation for the proper boundary conditions gives $n_e(x)$ in the $p$-region;

(c) from $n_e(x)$ find $dn_e/dx$ and employ this information to calculate the electron diffusion current at $x_p$;

(d) repeating steps (b) and (c) for the hole minority carriers in the $n$-region ($x \leqslant 0$) find the hole current at $x = 0$;

(e) the total junction current is obtained by adding the electron current at $x = x_p$ and the hole current at $x = 0$.

## 7.6 Thickness and capacitance of the junction barrier

In the preceding sections we were concerned with calculating the current-voltage relationship for an $n$-$p$ junction. In this discussion the potential barrier between the $n$- and $p$-regions played an important role. So far, however, nothing has been said about the thickness of the barrier layer. In the present section we shall discuss this problem on the basis of a simplified picture in which it is assumed that the transition from the $n$-region to the $p$-region takes place abruptly in a plane perpendicular to the $x$-axis going through the point $x = x_1$ (see Fig. 7.8). We shall assume that everywhere in the $n$-region the donor concentration is constant and equal to $N_d$ per m$^3$. Similarly we shall assume the acceptor concentration $N_a$ to be constant everywhere in the $p$-region. When the junction is in thermal equilibrium, the conventional potential (referring to a positive unit charge) across the junction varies as indicated in Fig. 7.8(b). We shall measure $x$ to the left, as indicated, and denote by $x = 0$ the point such that for $x \leqslant 0$ the potential may be considered as constant. The potential in the region $x \leqslant 0$ will be taken as zero. The potential in the region left of the point $x = x_2$ is also constant and presumably equal to the contact potential $V_0$ in thermal equilibrium. Given $N_d$, $N_a$ and $V_0$, what is the thickness of the barrier, $x_2$? In order to answer this question, we first remind the reader that in thermal equilibrium, electrons have been transferred from part of the $n$-region into the $p$-region, thus producing a positive space charge in

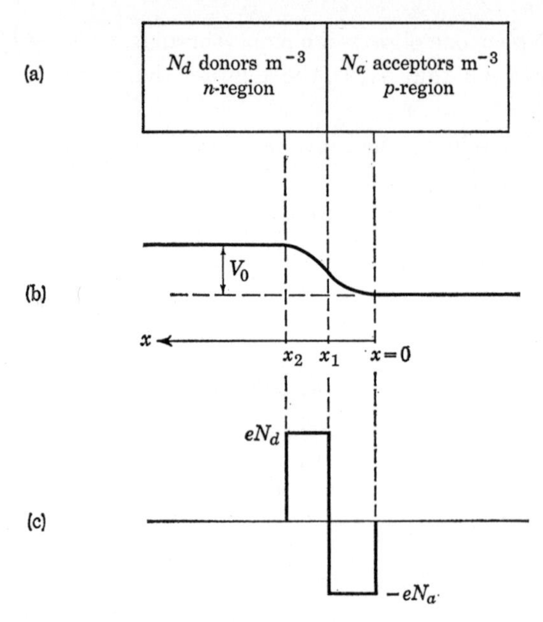


**Fig. 7.8.** (a) represents an $n$-$p$ junction which is assumed to change abruptly from $n$ to $p$ for $x = x_1$. In (b) the conventional potential has been indicated when the junction is in thermal equilibrium, corresponding to a contact potential $V_0$. In (c) the assumed space charge is represented.

the $n$-region resulting from uncompensated positive ionized donors. Similarly, holes have been transferred from the $p$-region into the $n$-region, leaving a net negative space charge in the $p$-region. Since the electron density in the region between $x_1$ and $x_2$ is small, we shall assume that the space charge density in that region is simply $eN_d$ as indicated in Fig. 7.8(c). Similarly, in the region between $x = 0$ and $x = x_1$ we shall assume a space charge density $-eN_a$. The potential is determined by the space charge through Poisson's equation. For the two regions we thus have the Poisson equations

$$\frac{d^2V_1}{dx^2} = \frac{eN_a}{\epsilon} \quad \text{for the region } 0 \leqslant x \leqslant x_1 \tag{7.43}$$

$$\frac{d^2V_2}{dx^2} = -\frac{eN_d}{\epsilon} \quad \text{for the region } x_1 \leqslant x \leqslant x_2 \tag{7.44}$$

Here, $\epsilon = \epsilon_0\epsilon_r$ where $\epsilon_0 = 8.854 \times 10^{-12}$ farad m$^{-1}$ and $\epsilon_r$ is the relative dielectric constant of the medium. Integrating expression (7.43) under the condition that in the point $x = 0$ $(dV_1/dx)$ is zero, we find

$$\frac{dV_1}{dx} = \frac{eN_a}{\epsilon}x \qquad 0 \leqslant x \leqslant x_1 \tag{7.45}$$

Integrating this equation, and remembering that we normalized the potential such that it is zero for $x = 0$, we find

$$V_1(x) = \frac{eN_a}{2\epsilon} x^2 \qquad 0 \leqslant x \leqslant x_1 \tag{7.46}$$

Hence, as we move to the left from the point $x = 0$, the potential rises quadratically with distance [see Fig. 7.8(b)].

The rest of the procedure runs as follows: when expression (7.44) is integrated twice, there will occur two integration constants in the general solution for $V_2(x)$. However, in the point $x = x_1$ the potential and the field strength obtained from the solution (7.46) must be the same as the corresponding quantities obtained by using the solution of (7.44) for the point $x = x_1$. This gives two equations from which the two integration constants can be found, so that $V_2(x)$ will then be known also. Finally, $V_2(x_2)$ can be put equal to the value $V_0$ which is assumed to be known, and from the resulting equation we shall be able to find $x_2$. The algebra involved in this procedure is shown below.

From (7.46) we obtain

$$V_1(x_1) = \frac{eN_a}{2\epsilon} x_1^2 \qquad \text{and} \qquad \left(\frac{dV_1}{dx}\right)_{x_1} = \frac{eN_a}{\epsilon} x_1 \tag{7.47}$$

Integrating (7.44) once and applying this to the point $x = x_1$ we obtain

$$\left(\frac{dV_2}{dx}\right)_{x_1} = -\frac{eN_d}{\epsilon} x_1 + A_1 \tag{7.48}$$

where $A_1$ is an integration constant which can be found from (7.48) and the second equation in (7.47) by equating the two derivatives. Hence, $A_1 = (ex_1/\epsilon)(N_d + N_a)$. Integrating (7.44) twice, we obtain the general solution

$$V_2(x) = -\frac{eN_d}{2\epsilon} x^2 + A_1 x + A_2 \tag{7.49}$$

where $A_1$ has been determined before and $A_2$ is a second integration constant. Applying (7.49) to the point $x = x_1$, substituting $A_1$, and making use of the fact that $V_2(x_1) = V_1(x_1)$ we obtain $A_2 = -(x_1^2 e/2\epsilon)(N_a + N_d)$. The potential $V_2$ in the point $x_2$ obtained by substituting for $A_1$ and $A_2$ in (7.49) is thus given by

$$V_2(x_2) = V_0 = -\frac{eN_d}{2\epsilon} x_2^2 + \frac{e}{\epsilon}(N_d + N_a)x_1 x_2 - \frac{e}{2\epsilon}(N_a + N_d)x_1^2 \tag{7.50}$$

Note that we cannot calculate $x_2$ from this equation unless $x_1$ can be expressed in terms of $x_2$. This can be done by realizing that at the point $x_2$, the field strength vanishes. Hence, in accordance with (7.49),

$$\left(\frac{dV_2}{dx}\right)_{x_2} = 0 = -\frac{eN_d}{\epsilon}x_2 + \frac{ex_1}{\epsilon}(N_d + N_a)$$

so that

$$x_1 = \frac{N_d}{N_d + N_a}x_2 \tag{7.51}$$

Substitution of $x_1$ into (7.50) finally leads to the equation

$$V_0 = \frac{e}{2\epsilon}\left(\frac{N_dN_a}{N_d + N_a}\right)x_2^2 \tag{7.52}$$

and consequently

$$x_2 = \left[\frac{2\epsilon V_0(N_d + N_a)}{eN_dN_a}\right]^{1/2} \tag{7.53}$$

Consider a typical example with $N_d \approx N_a \approx 10^{21}$ per m³ (see Table 7.1), $V_0 \cong 0.5$ volt, $\epsilon_0 = 8.85 \times 10^{-12}$ farad m$^{-1}$, $\epsilon_r \cong 16$ (for germanium) and $e = 1.6 \times 10^{-19}$ coulomb. We then find $x_2 \approx 10^{-6}$ m.

It is evident that this procedure may also be applied to the case of a biased junction. Thus, for a forward bias of $V$ volts, we would simply have to replace $V_0$ in equation (7.53) by $V_0 - V$, illustrating that the *thickness of the barrier decreases with increasing forward bias*. Similarly, *the thickness of the barrier increases with increasing reverse bias.*

From the electrical engineering point of view, the barrier layer is equivalent with a charged condenser. In fact, the total positive charge in the $n$-type region is equal to $Q = eN_d(x_2 - x_1)$ coulombs per m², and the tota negative charge in the $p$-type region is $-eN_ax_1$ coulombs per m²; because of relation (7.51) the positive and negative charges per m² have equal magnitudes. The effective capacitance $C$ per m² of the barrier layer follows from the equation

$$C = \frac{dQ}{dV_0} = \frac{d}{dV_0}[eN_d(x_2 - x_1)] = \frac{d}{dV_0}\left[ex_2\frac{N_dN_a}{N_d + N_a}\right] \tag{7.54}$$

where the last equality follows from (7.51). Substituting for $x_2$ from (7.53) in (7.54) and performing the differentiation one finds

$$C = \left[\frac{e\epsilon N_dN_a}{2V_0(N_d + N_a)}\right]^{1/2} = \frac{\epsilon}{x_2} \tag{7.55}$$

where the last equality follows from (7.53). Hence, the capacitance of the layer is the same as that of a parallel plate condenser with a distance $x_2$ between the plates and filled with a medium of dielectric constant $\epsilon = \epsilon_0\epsilon_r$. Since $x_2$ is of the order of $10^{-6}$ m in a typical case, the capacitance of the barrier layer is of the order of 150 microfarads per m².

## 7.7 The n-p-n junction transistor

In this section we shall discuss the physical processes which determine the action of a junction transistor. The structure of an *n-p-n* junction transistor is represented in Fig. 7.9; it consists of two *n*-type regions with a *p*-type region sandwiched between them. One of the *n-p* junctions is

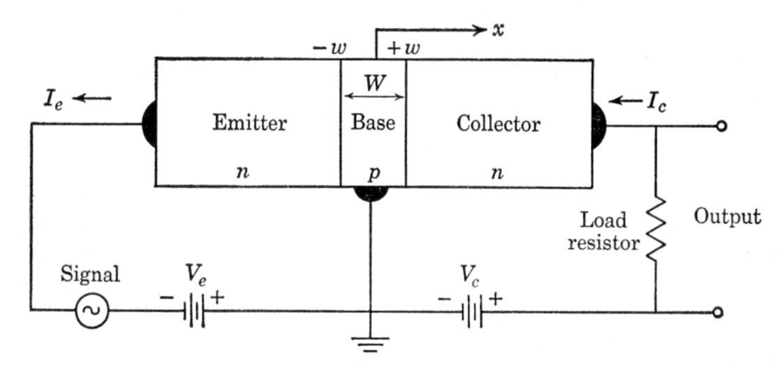


**Fig. 7.9.** Diagram of an *n-p-n* junction transistor with a base-to-ground connection. The emitter junction is biased in the forward direction, the collector junction in the reverse direction.

biased in the forward direction (emitter junction), the other in the reverse direction (collector junction). The *p*-region is called the base region; the two *n*-regions are referred to as the emitter and collector regions. A junction transistor may also be of the *p-n-p* type and the arguments given below apply equally well to it as to the *n-p-n* transistor under discussion. For a typical junction transistor, the characteristics describing the various regions are of the following orders of magnitude:

conductivity of the emitter region, $\sigma_e \approx 10^4$ ohm$^{-1}$ m$^{-1}$,
conductivity of the base region, $\sigma_b \approx 10^2$ ohm$^{-1}$ m$^{-1}$,
conductivity of the collector region, $\sigma_c \approx 10$ ohm$^{-1}$ m$^{-1}$,
diffusion length of minority carriers, $L \approx 10^{-3}$ m,
width of the base region, $W \approx 10^{-5}$ m ($\ll L$).

In Fig. 7.10 we have represented the electron potential in the three regions, as determined by the bottom of the conduction band, for the biased transistor. The magnitudes of the emitter bias, $V_e$, and of the collector bias, $V_c$, are indicated. Let us consider the currents flowing in the device in the absence of an a-c signal, but with the bias voltages applied as indi-

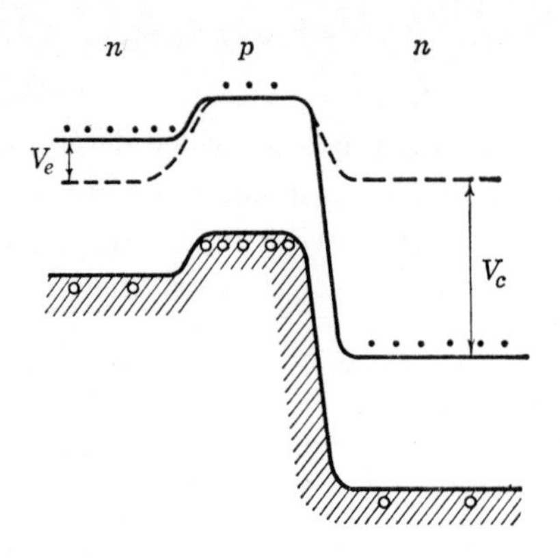


**Fig. 7.10.** Illustrating the energy band picture for the biased transistor.

cated. Since the emitter is biased negatively with respect to the base region, electrons are injected by the emitter region into the base region. Also, holes are injected by the base region into the emitter region. The total emitter current $I_e$ then consists of an electronic current, $I_{ee}$, and a hole current, $I_{eh}$, so that

$$I_e = I_{ee} + I_{eh} \tag{7.56}$$

In Fig. 7.9 the emitter current has been indicated as a conventional positive current leaving the external ohmic contact of the emitter region. Two conditions must be satisfied for good transistor action:

(a) nearly the whole emitter current should consist of electrons injected from the emitter $n$-region into the base $p$-region;

(b) nearly all electrons emitted by the emitter should travel through the base and be collected by the collector.

The reason for these requirements will become clear later, but we may state at this point that if these requirements are satisfied, one collects a current across the high resistance of the reverse-biased collector junction, whereas it is injected through the low resistance of the forward-biased emitter junction; this provides *power amplification.*

In connection with (a) the question can be raised as to what determines the ratio $I_{ee}/I_{eh}$? At first glance one may be tempted to employ expression (7.42), which would give $I_{ee}/I_{eh} = (\sigma_e/\sigma_b)(L_h/L_e) \cong \sigma_e/\sigma_b$. This answer is *incorrect,* however, because the base region has a width which is small compared to the diffusion length of minority carriers. This has the effect of increasing the ratio $I_{ee}/I_{eh}$ and one can in fact show that

$$I_{ee}/I_{eh} = (\sigma_e/\sigma_b)(L_h/W) \tag{7.57}$$

The reason for replacing the diffusion length of electrons in the base material $L_e$ by the width of the base region $W$ is the following: due to the electron injection by the emitter into the base region, the density of electrons in the base region at $x = -w$ (see Fig. 7.9) is raised to the value

$$n_e(-w) = n_{e0b}e^{eV_e/kT} \tag{7.58}$$

in accordance with (7.38); $n_{e0b}$ is the equilibrium density of electrons in the base region and $V_e$ is the magnitude of the emitter bias. Similarly, the density of electrons in the base region at the position $x = +w$ is reduced to the value

$$n_e(+w) = n_{e0b}e^{-eV_c/kT} \cong 0 \tag{7.59}$$

where $V_c$ is the magnitude of the collector voltage; since $eV_c \gg kT$, we have $n_e(+w) \cong 0$. Now, if we assume that the base region is so narrow that we may neglect recombination of electrons in the base region for the moment, the electron current in the base region due to diffusion from $x = -w$ to $x = +w$ is independent of $x$, and consequently $dn_e/dx$ must be constant in the base region. This is indicated in Fig. 7.11. The electron

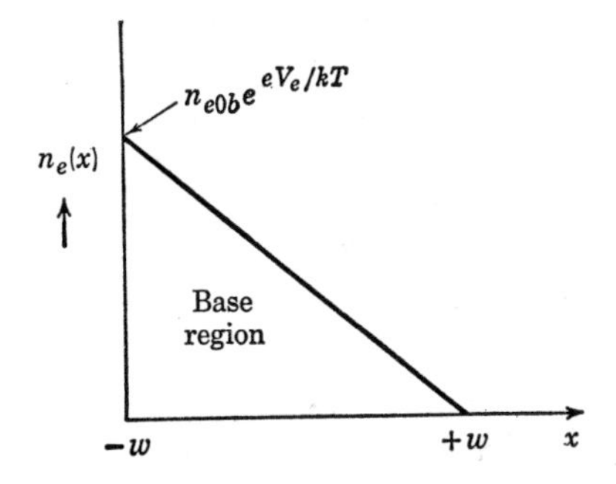


**Fig. 7.11.** The density of electrons in the base region of the *n-p-n* transistor varies linearly with $x$.

current resulting from the diffusion in the base region is then given approximately by

$$I_{ee} \cong -AeD_e\frac{dn_e}{dx} \cong \frac{eD_e}{W}n_{e0b}e^{eV_e/kT} \tag{7.60}$$

where $A$ is the cross section and $W = 2w$ is the width of the base region. Comparing this expression with (7.39) for $eV_e \gg kT$ we see that $L_e$ has been replaced by $W$. The hole current from the base to the emitter region, on the other hand, is not affected by the width of the base; these arguments thus show that (7.57) is correct.

The *emitter efficiency*, $\gamma$, may be defined as the fraction of the emitter current which is carried by electrons, i.e.,

$$\gamma = I_{ec}/I_e = I_{ee}/(I_{ee} + I_{eh}) \tag{7.61}$$

For the numerical values given in the beginning of this section we find from (7.57) that $I_{ee}/I_{eh} \cong 10^4$, so that $\gamma$ is nearly unity.

With regard to the requirement (b) mentioned before, we should consider what fraction of the electrons injected by the emitter into the base region actually arrives at the collector. If the base region has a width $W \ll L_e$, nearly all electrons injected into the base region will be collected by the collector. By solving the diffusion equation for electrons in the base region one can show that a fraction $\frac{1}{2}(W/L_e)^2$ of the electrons disappears by recombination as the electrons travel through the base region. We may thus introduce the *transport factor*, $\beta$, defined by

$$\beta = \frac{I_{ec}}{I_{ee}} \cong 1 - \frac{1}{2}\left(\frac{W}{L_e}\right)^2 \tag{7.62}$$

where $I_{ec}$ is the electron current collected at the collector.

The currents $I_{ee}$, $I_{eh}$, and $I_{ec}$ discussed so far are indicated schematically in Fig. (7.12). There is, of course, current flow between the base and col-

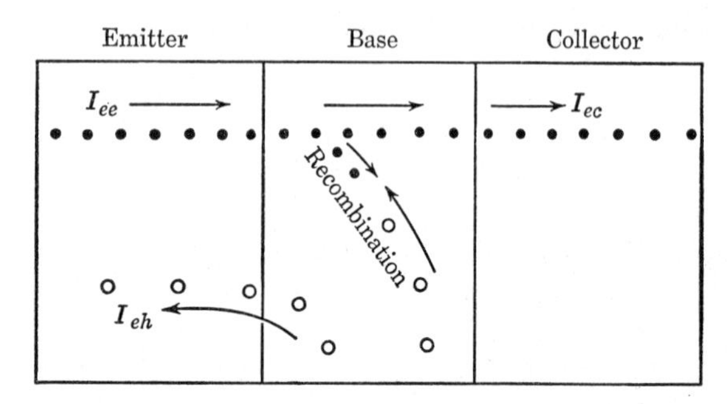


**Fig. 7.12.** Schematic representation of the most important electron and hole currents flowing in an *n-p-n* junction transistor. The dots represent electrons; the circles holes. The arrows indicate the direction of motion of the carriers, not the direction of conventional current flow, except for holes.

lector regions, also. Since the collector junction is biased in the reverse direction, these currents are very small compared to the currents $I_{ee}$ and $I_{ec}$, and we shall not discuss them here. In this approximation, then, we find from (7.61) and (7.62) for the ratio of the total collector current to the total emitter current

$$\alpha = \frac{I_c}{I_e} \cong \frac{I_{ec}}{I_e} = \beta\gamma \tag{7.63}$$

From these considerations we find the *approximate equivalent circuit* for the grounded-base junction transistor as indicated in Fig. 7.13. The resistance $r_c$ represents the collector junction resistance; it is large compared to $r_e$ because of the reverse bias of the collector. The capacitance $C_c$ is of importance only for a-c signals; it represents the capacitance of the collector junction barrier and, because it tends to short-circuit the

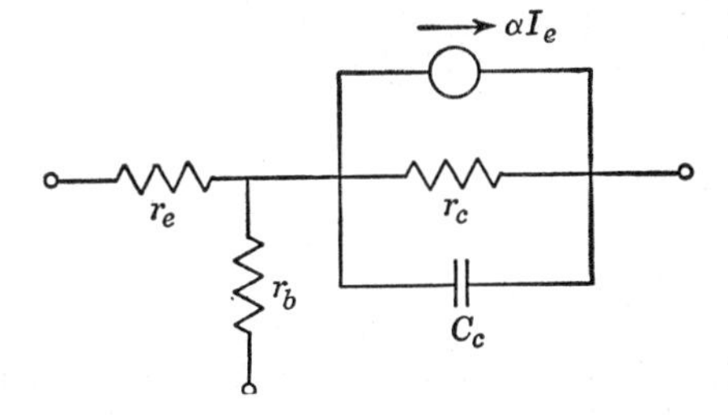


**Fig. 7.13.** Approximate equivalent circuit for the junction transistor.

collector resistance, it is one of the factors which limits high frequency response. The current generator of strength $\alpha I_e$ forms the active element of the transistor. The resistance $r_b$ gives the resistance produced by the base material.

For a typical transistor under operating bias conditions the values of these parameters are as follows:

$$r_e \cong 50 \text{ ohms} \qquad \alpha \cong 0.995$$
$$r_c \cong 5 \times 10^6 \text{ ohms} \qquad C_c \cong 20 \text{ micromicrofarads}$$
$$r_b \cong 200 \text{ ohms}$$

For a more detailed discussion of transistors and transistor circuits the reader is referred to the literature on the subject.

## References

A. W. Lo, R. O. Endres, J. Zawels, F. D. Waldhauer and C. C. Cheng, *Transistor Electronics*, Prentice-Hall, 1955.

R. D. Middlebrook, *An Introduction to Junction Transistor Theory*, Wiley, New York, 1957.

## Problems

**7.1** Indium antimonide (InSb) has an electron mobility of 6 $m^2$ $volt^{-1}$ $sec^{-1}$ and a hole mobility of 0.2 $m^2$ $volt^{-1}$ $sec^{-1}$. The highest room temperature (300°K) resistivity found to date is $2 \times 10^{-4}$ ohm m. Assuming the material is intrinsic, determine the intrinsic carrier density $n_i$ at room temperature.

**7.2** Gallium antimonide (GaSb) has an electron mobility of 0.3 $m^2$ $volt^{-1}$ $sec^{-1}$ and a hole mobility of 0.1 $m^2$ $volt^{-1}$ $sec^{-1}$. The highest resistivity of this material at 300°K found to date is $2 \times 10^{-3}$ ohm m. Assuming the material is intrinsic, calculate the intrinsic carrier density $n_i$ at 300°K. Compare the answer with that for indium antimonide in the preceding problem, and comment on whether this agrees with the fact that the forbidden energy gaps of GaSb and InSb are respectively 0.80 and 0.20 ev; what do you conclude?

**7.3** The electron and hole mobilities in a silicon single crystal at room temperature (300°K) are, respectively, 0.17 and 0.025 $m^2$ $volt^{-1}$ $sec^{-1}$. Find the diffusion coefficient of electrons and holes at this temperature.

**7.4** A $p$-type silicon crystal has a room temperature (300°K) resistivity of $9 \times 10^{-4}$ ohm m. What would be the Hall coefficient of the sample. Mobilities are given in the preceding problem.

**7.5** The lifetimes of carriers in a semiconductor in part are strongly determined by the condition of the surface. In one particular experiment, the lifetime of electrons at room temperature in an $n$-type germanium crystal with a ground surface is found to be 78 microseconds, whereas with an acid-etched surface it is 340 microseconds. Assuming an electron mobility of 0.36 $m^2$ $volt^{-1}$ $sec^{-1}$, find the diffusion length of the electrons in both specimens.

**7.6** The bulk $n$-region of an $n$-$p$ germanium junction has a resistivity at 300°K of $10^{-4}$ ohm m; the resistivity of the bulk $p$-region is $10^{-2}$ ohm m. Find the carrier concentrations from Table 7.1. Assuming the junction is in thermal equilibrium, find the potential drop across the junction, applying Boltzmann statistics to the carrier densities.

**7.7** The junction referred to in the preceding problem is subjected to a forward bias of 0.25 volts. Find the electron density in the $p$-region immediately adjacent to the junction (point $x_p$ in Fig. 7.6). Find also the electron current across the junction if the junction is circular with a radius of 0.15 mm. What is the hole current across the junction? The electron and hole mobilities are 0.36 and 0.17 $m^2$ $volt^{-1}$ $sec^{-1}$ respectively; assume a lifetime of 100 microseconds for both electrons and holes.

**7.8** For a $p$-$n$ germanium junction, let the conductivity of the bulk $p$-region be $10^4$ $ohm^{-1}$ $m^{-1}$ and that of the bulk $n$-region 100 $ohm^{-1}$ $m^{-1}$. In thermal equilibrium, the voltage drop across the barrier layer is 0.5 volt. Calculate the capacitance of the junction if its circular cross section has a diameter of 0.15 mm. Calculate also the capacitance for a reverse voltage of 3 volts. You may use Table 7.1 and assume that the junction is abrupt. The relative dielectric constant of the material is 16.

# Answers to Problems

**1.1** $2.69 \times 10^{25}$ m$^{-3}$
**1.2** appr. $3 \times 10^{12}$ m$^{-3}$
**1.3** 0.039 ev; appr. 2000 m sec$^{-1}$
**1.4** $5.93 \times 10^5$ m sec$^{-1}$; $1.38 \times 10^4$ m sec$^{-1}$
**1.5** $W_{\text{kin}} = 13.6$ ev; $W_{\text{pot}} = -27.2$ ev; $W_{\text{total}} = -13.6$ ev
**1.6** $W_1 = -13.6$; $W_2 = -3.4$; $W_3 = -1.51$; $W_4 = -0.85$ ev
$r_1 = 0.529$; $r_2 = 2.12$; $r_3 = 4.76$; $r_4 = 8.46$ angstroms
**1.7** $W_2 - W_1 = 10.2$ ev; $f = 2.47 \times 10^{15}$ sec$^{-1}$; in ultraviolet
**1.10** 600 volts
**1.15** $(5 - 4 - 2.9) = -1.9$ ev
**1.16** 2 for b-c-c.; 4 for f-c-c.; 1 for simple cubic lattice
**1.23** 7500 volts; 54.6 volts
**1.24** $8.5 \times 10^{28}$ m$^{-3}$

**2.1** For $r < R$, $D = 0$, $E = 0$, $V = Q/4\pi\epsilon_0 R$ volts
For $r > R$, $D = Q/4\pi r^2$ coulomb m$^{-2}$, $E = Q/4\pi\epsilon_0 r^2$ volt m$^{-1}$, $V = Q/4\pi\epsilon_0 r$ volts
**2.2** For $r < R$, $D = Qr/4\pi R^3$ coulomb m$^{-2}$, $E = Qr/4\pi\epsilon_0 R^3$ volt m$^{-1}$, $V = (Q/8\pi\epsilon_0 R)(3 - r^2/R^2)$ volts
For $r > R$, $D = Q/4\pi r^2$ coulomb m$^{-2}$, $E = Q/4\pi\epsilon_0 r^2$ volt m$^{-1}$, $V = Q/4\pi\epsilon_0 r$ volts
**2.3** $C = 4\pi\epsilon_0 R$ farads; stored energy $= Q^2/8\pi\epsilon_0 R$ joules
**2.4** $Q = 2\pi\epsilon_0\epsilon_r V/[\ln(R_2/R_1)]$ coulomb m$^{-1}$;
$C = 2\pi\epsilon_0\epsilon_r/[\ln (R_2 R_1)]$ farad m$^{-1}$
**2.5** $C = 5.7$ microfarads
**2.7** For $TiO_2$ dielectric, 0.50 and 0.495 joule; for mica, 0.50 and 0.407 joule
**2.9** $Q$ coulomb m, along positive $x$-axis
**2.10** $E = 2.82 \times 10^7$ volt m$^{-1}$; $\mu = 2.5 \times 10^{-11}$ coulomb m
**2.11** $\alpha = Q^2/f$ farad m$^2$
**2.12** $\epsilon_r = 1.000435$
**2.13** $\alpha = 3\pi\epsilon_0 r_1^3$ farad m$^2$

**2.15** $\mu = 1.44 \times 10^{-31}$ coulomb m = 0.043 Debye units; attractive force = $4.2 \times 10^{-13}$ newton
**2.16** $2.44 \times 10^{25}$ $m^{-3}$; $\mu = 0.95$ Debye units; $\alpha = 1.6 \times 10^{-39}$ farad $m^2$
**2.17** $\mu_{ind} = \alpha\mu/2\pi\epsilon_0 a^3$
**2.18** $\mu_{ind} = -\alpha\mu/4\pi\epsilon_0 a^3$
**2.19** $E_i = E/(1 - \alpha/2\pi\epsilon_0 a^3)$ volt $m^{-1}$ for each atom; $E_i/E = 1.029$
**2.20** $E_i = E(1 + \alpha/4\pi\epsilon_0 a^3)$ volt $m^{-1}$ for each atom; $E_i/E = 0.986$
**2.22** $E_i/E = 1.6$
**2.23** The ferroelectric configuration

**3.4** $C = 200$ micromicrofarads; $R = 2 \times 10^6$ ohms; $W = 10^{-6}$ joule $sec^{-1}$

**4.1** $F_y = -5$ newtons
**4.2** $5 \times 10^{-9}$ weber $m^{-2}$
**4.4** $10^{-4}$ newton (attr.)
**4.5** $\mathbf{F} = -e(\mathbf{E} + \mathbf{v} \times \mathbf{B})$
**4.9** $B = 1.257$ weber $m^{-2}$; $M = 5$ ampere $m^{-1}$
**4.10** $B = 1.257$ weber $m^{-2}$; $M = 1400$ and 5000 ampere $m^{-1}$
**4.12** 2.2 Bohr magnetons per atom
**4.13** $2.8 \times 10^{-3}$ and $\sim 0.2$ Bohr magnetons

**5.1** $\rho = 1.72 \times 10^{-8}$ ohm m
**5.2** $v_{drift} = 0.7$ m $sec^{-1}$; $\mu = 7 \times 10^{-3}$ $m^2$ $volt^{-1}$ $sec^{-1}$; $\tau = 4 \times 10^{-14}$ sec
**5.3** $v_F = 1.39 \times 10^6$ m $sec^{-1}$; $\lambda = 556$ angstroms
**5.4** $\rho = 5.8 \times 10^{-8}$ ohm m; increase 222 percent
**5.5** $\rho_i = 0.97 \times 10^{-6}$ ohm m
**5.7** $F(+\,0.1 \text{ ev}) = 0.018$; $F(-\,0.1 \text{ ev}) = 0.982$
**5.9** $\rho(300°) = 1.866 \times 10^{-8}$ ohm m; $\rho(4°) = 0.306 \times 10^{-8}$ ohm m; 0.28 percent for the alloy and 0.33 percent for pure Cu
**5.10** $\Delta v_x = 0.476$ m $sec^{-1}$; $2.05 \times 10^{-31}$ joule
**5.11** $1.5 \times 10^8$ joules
**5.12** $K = 432$ and 5.25 watt $m^{-1}$ degree $^{-1}$

**6.1** 8
**6.2** $N = 5.00 \times 10^{28}$ $m^{-3}$ for Si and $4.52 \times 10^{28}$ $m^{-3}$ for Ge
**6.4** $n_i = 2.38 \times 10^{19}$ $m^{-3}$
**6.5** $1.1 \times 10^{16}$ $m^{-3}$
**6.6** $n_h = 4.52 \times 10^{24}$ $m^{-3}$; $n_e = 1.25 \times 10^{14}$ $m^{-3}$; $\rho = 0.36 \times 10^{-5}$ ohm m
**6.7** $n_h = 2 \times 10^{22}$ $m^{-3}$; $\mu_h = 0.035$ $m^2$ $volt^{-1}$ $sec^{-1}$ if formula (6.24) is used
**6.8** 1.83 millivolt
**6.9** 1.1 degree

**7.1** $n_i = 0.5 \times 10^{22}$ $m^{-3}$
**7.2** $n_i = 0.78 \times 10^{22}$

**7.3** $D_e = 0.0044$ m² sec$^{-1}$; $D_h = 0.00091$ m² sec$^{-1}$
**7.4** $R_H = 3.7 \times 10^{-5}$ coulomb$^{-1}$ m$^{-3}$ $\lambda$ if formula (6.24) is used
**7.5** $L = 0.85$ and 1.77 mm
**7.6** $V_0 = 0.5$ volt
**7.7** $n_e(x_p) = 2.7 \times 10^{21}$ m$^{-3}$; $I_e = 0.3$ milliamperes; $I_h = 0.004$ milliamperes
**7.8** 3.5 micromicrofarads; 1.3 micromicrofarads

**17.** A conical water glass is to be made so that when a heavy sphere 2 inches in diameter is placed inside and the glass is filled with water, the sphere will barely be submerged. Find the semivertical angle of the cone if the volume of the glass is the least possible.

**18.** (a) For what values of $x$ is $f(x) = (60x - 46x^2 + 12x^3 - x^4)^{1/2}$ defined? (b) For what values of $x$ is $f$ differentiable? (c) Find the absolute maximum and the absolute minimum of $f$. (d) Are there any two-sided relative extrema which are not absolute extrema?

**19.** A ball is tossed straight up. The sun is setting, and the horizontal rays throw the shadow of the ball onto a nearby hemispherical dome, of radius 18 feet. The ball is thrown so that it rises exactly to the height of the top of the dome. (a) Find the speed of the shadow along the surface of the dome as a function of $t$ when $t \geq 0$, if $t = 0$ is taken as the instant at which the ball reaches its highest point. (b) Evaluate at $t = 0$, and note the surprising character of the result.

**20.** A water glass has the shape of a cone of altitude $h$ and semivertical angle $\phi$. The glass is filled with water, and into it is carefully lowered a spherical ball of such size as to cause the greatest possible overflow. Find the radius of the ball.

# *Index*